Core Principles of Maritime Navigation

Core Principles of Maritime Navigation introduces the main navigation concepts required by trainees and professionals involved in maritime operations. The book covers a wide range of subjects from wind, waves and depths to navigational marks and buoys, navigational techniques and equipment, passage planning, duties of the Officer of the Watch, berthing and anchorage. It is not intended to be a technical publication; rather, it aims to introduce core ideas and concepts in an accessible way for general readers, university students, cadets and for the continuous professional development of experienced seafarers.

Alexander Arnfinn Olsen is a Senior Learning and Development Consultant for a leading UK maritime consultancy, instructional designer and freelance writer. He is the author of *Introduction to Ship Operations and Onboard Safety* (Routledge, 2022).

T0246314

Core Principles of Maritime Navigation

Alexander Arnfinn Olsen

Routledge
Taylor & Francis Group

LONDON AND NEW YORK

Cover image: © Tom Fisk – courtesy of Pexels. Aerial shot of cargo ship, North Jakarta, Indonesia.

First published 2023
by Routledge
4 Park Square, Milton Park, Abingdon, Oxon OX14 4RN

and by Routledge
605 Third Avenue, New York, NY 10158

Routledge is an imprint of the Taylor & Francis Group, an informa business

British Library Cataloguing-in-Publication Data
A catalogue record for this book is available from the British Library

Library of Congress Cataloging-in-Publication Data
Names: Olsen, Alexander Arnfinn, author.
Title: Core principles of maritime navigation / Alexander Arnfinn Olsen.
Description: Abingdon, Oxon ; New York, NY : Routledge, [2022] | Includes bibliographical references and index. | Identifiers: LCCN 2022001339 (print) | LCCN 2022001340 (ebook) | ISBN 9781032271392 (hardback) | ISBN 9781032271385 (paperback) | ISBN 9781003291534 (ebook)
Subjects: LCSH: Navigation--Handbooks, manuals, etc. | Merchant marine--Handbooks, manuals, etc. | Seamanship--Handbooks, manuals, etc.
Classification: LCC VK155 .O468 2022 (print) | LCC VK155 (ebook) | DDC 623.89--dc23/eng/20220121
LC record available at https://lccn.loc.gov/2022001339
LC ebook record available at https://lccn.loc.gov/2022001340

ISBN: 978-1-032-27139-2 (hbk)
ISBN: 978-1-032-27138-5 (pbk)
ISBN: 978-1-003-29153-4 (ebk)

DOI: 10.1201/9781003291534

Typeset in Bembo
by SPi Technologies India Pvt Ltd (Straive)

Contents

vi *Contents*

Figures

Tables

Image Attributions

My grateful thanks go to the following individuals who have made their photographic materials available to the public. If I have omitted to attribute anyone, or to do so correctly, this is an accidental oversight, for which I offer my sincere apologies.

Abbreviations

AC	Alternating Current
AIO	Admiralty Information Overlay
AIS	Automatic Identification System
AIS–SART	Automatic Identification System – Search and Rescue Transponder
ALRS	Admiralty List of Radio Signals
ARPA	Automatic Radar Plotting Aid
ASD	Azimuth Stern Drive
AtoN	Aid to Navigation
BNWAS	Bridge Navigational Watch and Alarm System
CALDOVREP	Calais Dover Strait (English Channel)
CATZOC	Categories of Zones of Confidence
CBM	Conventional Buoy Mooring
COG	Course over Ground
COLREGS	International Regulations for Preventing Collisions at Sea, 1972
COPREP	Off the Coast of Portugal
COSWP	Code of Safe Working Practices for Merchant Seafarers
CPA	Closest Point of Approach
CPP	Constant Pitch Propeller
DC	Direct Current
DNVGL	Det Norske Veritas Germinischer Lloyd
DR	Dead Reckoning
DSC	Digital Selective Caller
EBL	Electronic Bearing Line
ECDIS	Electronic Chart and Display Information Systems
ENC	Electronic Navigational Chart
EP	Estimated Position
EPIRB	Emergency Position Indicating Radio Beacon
ERBL	Electronic Range and Bearing Line
ETA	Estimated Time of Arrival
FINREP	Off the Coast of Finland
FPSO	Floating Production, Storage, and Offloading
FT	Feet
GIBREP	Strait of Gibraltar
GIGO	Garbage In Garbage Out

GMDSS	Global Maritime Distress and Safety System
GOC	General Operator's Certificate
GPS	Global Positioning System
GRT	Gross Registered Tonnage
HSA	Horizontal Sextant Angle
HMCS	HM Canadian Ship
IALA	International Association of Marine Aids to Navigation and Lighthouse Authorities
IBS	Integrated Bridge System
ICZ	Intertropical Convergence Zone
IHO	International Hydrographic Organisation
IMDG	International Code for the Maritime Transport of Dangerous Goods in Packaged Form
IMO	International Maritime Organisation
ISM	International Safety Management Code
ITU	International Telecommunications Union
KGs	Kilograms (metric)
kHz	Kilohertz
LBS	Pounds (imperial)
LNG	Liquid Natural Gas
LOA	Length Overall
LOP	Lines of Position
LPG	Liquid Petroleum Gas
LRIT	Long Range Identification and Tracking System
LUT	Local User Terminal
LWL	Length, Waterline
MBL	Maximum Breaking Load
Metareas	Meteorological Areas
MHz	Megahertz
MMI	Man Machine Interface
MMSI	Maritime Mobile Service Identity
MOB	Man overboard
MRCC	Maritime Rescue Coordination Centre
MSI	Maritime Safety Information
MV	Motor Vessel
Navareas	Navigational Areas
NAVTEX	Navigational Telex
NBDP	Narrow Band Direct Printing
NC	New Chart
NE	New Edition
NM	Nautical Mile
NP	Nautical Publication
NTM	Notices to Mariners
NUC	Not Under Command
OCIMF	Oil Companies International Marine Forum
ODAS	Ocean Data Acquisition System

OiC	Officer in Charge
OOW	Officer of the Watch
PI	Parallel Indexing
PLB	Personal Location Beacons
PLEM	Pipeline End and Manifold
PMR	Portable Marine Radio
PPE	Personal Protective Equipment
PPI	Plan Position Indicator
RACONS	Radar Beacons
RAI	Rudder Angle Indicator
RAMARK	Radar Markers
RMS	Royal Mail Ship
SAR	Search and Rescue
SART	Search and Rescue Transponder
SCAMIN	Scale Minimum
SENC	System Electronic Navigational Chart
SMS	Safety Management System
SOG	Speed over Ground
SOLAS	International Convention for the Safety of Life at Sea, 1974
SPM	Single Point Mooring
SSM	Ship to Ship Mooring
STDMA	Self Organising Time Division Multiple Access
STW	Speed Through Water
TCD	Turning Circle Diameter
TCPA	Time to Closest Point of Approach
TEU	Twenty-foot Equivalent
THD	Transmitting Heading Device
TPNM	Temporary and Preliminary Notices to Mariners
TRS	Tropical Revolving Storms
TSS	Traffic Separation Schemes
UIN	Unique Identifier Number
UKC	Under Keel Clearance
UKHO	United Kingdom Hydrographic Office
USCG	US Coastguard
USS	United States Ship
UTC	Coordinated Universal Time
VDR	Voyage Data Recorder
VHF	Very High Frequency
VMS	Vessel Monitoring System
VPP	Variable Pitch Propeller
VRM	Variable Range Marker
VSA	Vertical Sextant Angle
VSP	Voith Schneider Propeller
VTS	Vessel Tracking System
VTS	Vessel Traffic Services
WETREP	Western European Tanker Reporting System

WGS	World Geodetic System, 1984
WWNWS	Worldwide Navigational Warning Service
ZOC	Zones of Confidence

Preface

The navigating or deck officer must be extremely cautious when steering the vessel from its course no matter where the vessel might be – in the mid-sea, crossing a channel, or entering or leaving port. The bridge team should be sufficiently capable to sail the vessel safely in all kinds of waters and weather. This, of course, comes with experience and the development of skills. Simply knowing how to sail the vessel is very different from being able to sail the vessel.

To that end, this book seeks to provide the reader with the first half of this equation: the knowledge part. Naturally, navigation is a vast and complex subject area and there is simply too much to be able to safely condense into one book. What I have tried to do, therefore, is to provide the reader with some direction; a summary of some of the key concepts and themes that concern safe navigation.

— **Alexander Arnfinn Olsen**
Southampton, April 2022

Acknowledgements

The Author would like to personally thank everyone involved in putting together this book, with especial gratitude to Fidaa Karkori, without whose support this book would not have been possible, and to the publishing team at Routledge.

Author's note

Like all modern professions, the maritime industry uses a wide range of terminology and a rich vocabulary built up over the course of centuries. The reader may notice that some terms have been used interchangeably – such as ship and vessel – please be assured, this is perfectly correct and proper. I would also like to draw the reader's attention to the use of the terms ship's navigator, navigating officer, and Officer of the Watch. Although the role of ship's navigator is technically a responsibility delegated to the second officer, all deck officers, when on watch, are in effect navigating the ship. It stands to reason therefore they should be referred to as ship's navigators. The same principle applies to the use of navigating officer. When a deck officer is on watch, i.e., responsible for the safe navigation of the vessel, they assume the role of Officer of the Watch. This is usually abbreviated to OOW.

1   Wind, depth and waves

One of the many natural factors the ship's navigator must be conscious of when in command of a ship's steering is the effect of wind. Vessels such as container ships and roll-on, roll-off ships have large freeboards. This means they are more affected by the effect of wind than ships with lower freeboards, such as tankers and gas carriers. This exposed area of the ship is called the windage area, as it is here that the wind has the most prominent effect. The effect that wind has on a vessel is dependent on the location the wind assails the vessel's freeboard, and on the draught condition of the vessel. For example, a wind with a force of 3–4 on the Beaufort Scale will have a similar effect on a ship in light conditions as a wind with a force of 7–8 on the Beaufort Scale when the vessel is down to her marks. This means wind direction is important for the ship's navigator. When a vessel is making way at slow speed, manoeuvring, or when sailing near the coastline, wind direction is comparatively easy to establish. It is not nearly as easy to discern when the vessel is out in open sea. The reason being the perceived direction of the wind when standing on deck is relative to the position of the person making the judgement. This is because the true direction of the wind is masked by the vessel's course. Because manually determining wind direction is so difficult, modern ships are fitted with a piece of equipment called the gyro compass. The gyro compass is linked to a set of repeater compasses via a transmission system. Within the compass is a fast-spinning rotor that weighs between 0.56 kgs (1.25 lbs) and 24 kgs (55 lbs). This is driven thousands of revolutions per minute by an electric motor. Connected to the motor is the spinning wheel, which is called a gyroscope. It is the gyroscope which provides the ship's navigator with the true wind direction.

It is very important for the ship's navigator to account for wind direction when steering the vessel. Steering into the wind, for example, will increase drag on the vessel. This, in turn, will reduce the efficiency of the vessel's engines. This means the vessel will consume more fuel than necessary. Strong winds also generally equate to increased wave activity. Sailing head-on into opposing winds will not only make for a very uncomfortable condition onboard but will also result in reduced efficiency. Therefore, the ship's navigator is always strongly advised to sail with the wind wherever possible. There are many different conditions where a vessel will interact with wind. Understanding each of these conditions and the way they influence the ship's behaviour is a key skill experienced deck officers develop

DOI: 10.1201/9781003291534-1

over time. Below are some of the main conditions where wind can influence ship behaviour. Although this list is not exhaustive, it is nevertheless important to understand some of the most common ways winds interact with ships.

Underway with the wind from right astern

When the wind is blowing from right astern, steering the vessel becomes easier. In the event of a head wind, however, the stern part of the vessel will develop a tendency to pay off on either side. This is a difficult condition to tackle and getting the vessel back on course is invariably no easy task. Effects such as this are more often experienced on vessels where the accommodation superstructure is situated closer towards the aft section. It should be noted, however, that when the wind is right astern, it offers no braking effect for the vessel. This is important to remember, especially when berthing. Given the choice to berth with a head wind and a wind from right astern, it is always preferable to berth against a head wind as this provides some natural braking moment.

Underway with wind from abeam

When the vessel is underway with the wind flowing from abeam, the steering of the vessel is not usually affected. That said, depending on the strength of the wind, the vessel may develop a sideways drift caused by leeway. This must be accounted for and corrected by the ship's navigator.

Underway with the wind on the bow

In conditions where the wind is blowing in from the bow, the effect on the vessel's stem is more pronounced. This will cause the ship's head to swing away from the wind in a leeward direction. To counteract this, the weather helm (i.e., the helm facing the wind) will need continuous steering adjustment.

Underway with the wind on the quarter

When the wind is pushing the ship's stern away to the leeward side, the stern tends to swing towards the leeward. To counteract this, the vessel must be steered towards the wind and given a lee helm.

Under sternway

When the ship is astern it rarely does so at speed. Also, when going astern, most ships tend to develop a natural swing towards starboard. This means the effect that wind has on ship behaviour is slightly more complex. In a ballast condition where the wind catches the bow, which is often the case, the stern is pulled into the wind. This effect can be quite rapid and pronounced. Special attention must therefore be paid when manoeuvring to anchor or when berthing. All ships turn around an imaginary reference called a pivoting point. This point is fixed and can be used by the ship's navigator to aid in manoeuvring the ship. This also means the ship can harness the power of wind as a brake, for supporting a tight turn, or for helping the ship to manoeuvre relatively easily if the wind remains about two to three points on the bow.

Ocean currents

Thus far we have discussed how the ship's navigator can use the forces of wind to aid the safe manoeuvring of their ship. But we should also factor in the effects of ocean currents. Indeed, ocean currents play a very important role in ship stability, especially when the ship is under the effect of onshore winds, near offshore platforms, when making way through narrow or congested channels, on the open sea, and when in harbours or inland waters. When the current has a constant strength and direction, the ship's handling becomes much easier. Such conditions, however, only really exist in comparatively narrow channels and rivers. This means the navigator must account for different current streams that may exist in a relatively small area. Although currents do have an impact on ship handling, when compared to the effect of wind, currents are far more predictable. Even in open waters, when the ship is approaching a rig or a mooring buoy, due allowance must be given to the effect of current for a safer manoeuvre. Allowing the current to flow from the ships ahead will reduce the ship's speed over ground, improve the ship's rudder response, and provide more time to assess and correct for developing situations.

Ship's squat

When a ship proceeds through water, it pushes the water ahead. This volume of water returns down the sides of the ship and down and under the keel. The streamlines of return flow are sped up as the ship passes over, which causes a drop in pressure. This in turn results in the ship dropping vertically into the water. When the ship drops vertically in water, it trims both forward and aft. This decrease in static under keel clearance (ukc), both forward and aft, is called ship's squat. This is not so much of a concern in open sea, where the depth of water is considerable, but in shallower waters, squat can cause considerable alarm. If the ship moves forward at great speed in shallow waters, especially where the keel clearance is as little as 1 to 1.5 metres (3.2–4.9 ft), the risk of running aground at the bow or stern is greatly increased. Subsequently, the navigator must always pay attention to the ship's speed and ukc when making way through shallower waters. This is because the main factor that determines the ship's squat is speed. In essence, squat can be roughly calculated by squaring the ship's speed. Another factor to consider is the blockage factor. This is defined at the immersed cross-section of the ship's midsection divided by the cross-section of the body of water. The blockage factor ranges from about 8.25b for super tankers to about 9.50b for general cargo ships and 11.25b for container ships. The presence of other vessels in the vicinity of a narrow channel will also affect squat; so much so, that squats can double in value as the ship passes by or across the other vessel.

Shallow waters

Shallow waters present many hazards for navigators, the most obvious one being the risk of running aground. For this reason, navigational charts clearly indicate water depths at strategic positions. This enables the passage planner to chart a route through the most appropriate depths. But there are also other ways of determining whether the ship has entered shallow waters. These methods are important as navigational charts are not infallible and may contain errors; or, indeed, the ship may have been pushed off-course by strong winds or currents. First and foremost, wave generation from the bottom of the ship increases in shallower water, especially forwards towards the bow. The ship will often become sluggish and harder to manoeuvre. Draught indicators or echo-sounders will indicate changes in the end draughts. Propeller rpm will decrease. If the ship is in open water conditions, i.e., without breadth restrictions, this decrease may be 15% of the service rpm in deep water depths. If the ship is confined to a narrow channel, the decrease in rpm may be as high as 20% of normal service rpm. The drop in propeller rpm will inevitably lead to a fall in speed. If the ship is in open water conditions, this fall in speed can be as high as 35%. In confined channels, it can be as high as 75%. It is very common for the ship to start vibrating suddenly. This is because of the water effects causing the natural hull frequency to become resonant with other vessel-related frequencies. Any rolling, pitching, and heaving motions will reduce as the ship moves from deep water to shallow water conditions. This is caused by the cushioning effects produced by the narrow layer of water beneath the ship's keel. A reduction in depth between the ship's hull and the seabed will likely cause mud clouds to develop as the turbulence caused by the ship passing overhead disturbs the top layers of sediment. The Turning Circle Diameter (TCD) also increases in shallower waters; sometimes by as much as 100%. In addition, stopping distances and stopping times also increase when compared to deep water conditions. Finally, the effectiveness of the rudder helm decreases, making the ship feel heavy and unresponsive.

How waves are formed

As we discussed earlier, there are several types of waves, and the forces behind them are also different. The most common cause of ocean waves is wind. Wind-driven waves, also referred to as surface waves, are formed by the friction between the water's surface and the wind. When the wind is blowing on the sea, the surface exerts gravitational force on the bottom layer of the wind. This, in turn, exerts pull on the layers above until it reaches the top-most layer. With the gravitational pull being different at each layer, the wind moves at different speeds. The top-most layer tumbles, forming a circular motion. This creates downward pressure at the front and upward pressure at the rear of the surface, creating a wave. Wind is not the only cause of waves. Tidal waves, for example, are caused by the gravitational pull of the Sun and the Moon as the Earth rotates. It is important to note that a tidal wave is a shallow water wave, and not a tsunami, which is a different phenomenon altogether. Whereas the aforementioned waves are not dangerous in terms of their impact, there are hazardous wave formations that navigators need to be aware of.

This includes tsunamis. Tsunamis and other forms of hazardous waves are generally caused by severe weather conditions such as hurricanes, typhoons, tornados, and other natural disasters, including earthquakes, landslides, and volcanic eruptions.

Waves are essentially disturbances or oscillations on the surface of the water, which can be formed in all water bodies including oceans, seas, lakes, and rivers. Although waves stem from an external force, they are actually a restoring force which counters the effect of the disturbance. Put simpler, waves are the manifestation of kinetic energy passing through a body of water. This energy makes the water react in a circular motion. If we observe a boat encountering a wave, for example, we will note the wave pushes the boat upwards and forwards, swirls it round, then returns the boat to its original position. What this shows us is that although the water is moving, it is not actually travelling any distance. Some might argue that they have clearly seen waves move forward. This is most apparent along a shoreline. What is actually happening is an optical illusion. The inclined edge of the beach offers resistance and slows down the lower portion of the wave. This creates an imbalance which forces the upper portion of the wave – called the crest – to topple forward. It is this which we see at the beach, and which provides us with the illusion that the wave is moving forward. Having now established that waves represent the movement of energy, the obvious question is: where does this energy come from? While mild winds blowing over the surface of the water may create small surface waves, extreme weather conditions like hurricanes and cyclones produce strong winds and often create huge waves. These waves present a major hazard to shipping. Various other natural incidents, such as underwater earthquakes, landslides, and volcanic eruptions, can also generate massive releases of energy which in turn create huge waves called tsunamis. These are extremely destructive to coastal regions and have a vastly detrimental impact on marine ecology and human habitations. So, in summary, we can classify waves according to their formation, source of energy, and their behaviour. Now that we have covered the basics of wave formation, we can briefly turn our attention to the main types of waves.

Types of waves

There are many different types of waves, and each has their own specific and unique characteristics. We don't necessarily want to dwell too much on these, so we will turn our attention to some of the main wave types.

(1) *Breaking waves*. Breaking waves, as the name suggests, are formed when a wave collapses in on itself. These are most commonly observed on a coastline as wave heights are normally amplified in shallower water areas. When the wave approaches the shore, its profile is modified by the resistance of the sloping seabed. The seabed obstructs the motion of the base (or trough) of the wave, while the top part (or crest) continues to move at its usual speed. Subsequently, the wave begins to lean forward as it gradually reaches the shore. At a point where the steepness ratio of the wave reaches 1:7 the crest outruns the slow-moving trough, and the entire profile of the wave collapses onto itself, thereby, forming a breaking wave.

(2) *Spilling waves* or mushy waves. These form at gentle inclines on the ocean floor. If the shoreline is gently sloping, the energy of the wave is gradually expelled. This causes the crest to spill forming mild waves. These waves take more time to break compared to most other wave types.

(3) *Plunging waves.* When waves pass over a steeply inclined or rugged ocean floor, the crest of the wave curls and traps a pocket of air beneath it. As a result, the wave 'explodes' as it reaches a steeper gradient. The wave's energy is dissipated much quicker and over a much shorter distance. This causes the wave to plunge into the void space vacated by the trapped air. Plunging waves are common during offshore winds. They are extremely dangerous for surfers and beachgoers, and due to the vast amount of energy expelled, are responsible for causing the majority of coastal erosion.

(4) *Surging waves.* These are produced when huge swells reach shorelines with steep profiles. They travel at high speed and have no crest. Although they appear innocuous to the untrained eye, they have a strong backwash (pulling or sucking effect) which can drag people and animals under the water's surface and potentially out to sea. Surging waves are all the more dangerous as they do not break like other waves, and so are more difficult to identify.

(5) *Collapsing waves.* These are an unusual combination of plunging and surging waves. As the wave surges forward, the bottom profile of the wave aligns vertically with the crest. This then breaks completely as the wave loses energy and collapses, turning the wave into white water.

(6) *Deep-water waves.* As the name suggests, deep-water waves are formed where the water depth is significantly deeper compared with the other wave types. Importantly, there is no shoreline to provide any resistance to their motion. Technically, deep-water waves are formed where the depth of water is more than half the wavelength of the wave itself. The speed of the wave is measured as a function of the wavelength. In other words, waves with a longer wavelength travel at greater speeds compared to waves with shorter wavelengths. This means that multiple waves of differing wavelengths superimpose themselves on one another to form a single combined wave. Deep-water waves are long and travel in straight lines. They have sufficient energy to traverse longer distances compared to smaller coastal waves. The main source of deep-water waves is wind energy, which can be either local or distant. These are also known as stokes waves or short waves.

(7) *Shallow water waves.* The opposite of the deep-water wave is the shallow water wave. These waves are formed when the depth of the water is less than 1/20 of the wavelength of the wave. Unlike deep-water waves, the speed of the wave has nothing to do with the wavelength of the wave and everything to do with the depth of water. This means that waves in shallow waters move faster than waves in deeper waters. More specifically, the speed is equal to the square root of the depth of the water and the rate of acceleration is comparable to the force of gravity. To get scientific about it, we can use the following formula:

$$\text{speed} = \sqrt{(g.\text{depth})} \left(g = \text{gravitational constant, } 9.8\,\text{m/s}^2; \text{D} = \text{depth in metres} \right)$$

These are also known as Lagrangian waves or long waves.

(8) *Tidal wave.* These are not to be confused with tsunamis and are caused by the astronomical effect of the Sun and the Moon exerting gravitational pull on the Earth. As the Earth rotates, the Sun and the Moon pull the water in opposite directions, giving rise to high and low tides. In any 24-hour period, this occurs twice at 12-hourly intervals.

(9) *Tsunami.* These are often incorrectly called tidal waves. The word 'tsunami' originates from Japan and is a combination of 'tsu', which means harbour, and 'nami', which means wave. The majority (approximately 80%) of tsunamis or harbour waves are caused by large-scale underwater earthquakes. The remaining 20% are generated by underwater landslides, volcanic eruptions, and interstellar incidents such as meteor impacts. Because of the massive force which produces them, tsunamis travel at high velocity and are extremely dangerous and devastating. Interestingly, tsunamis are considered shallow water waves. This is because the typical tsunami wavelength is several hundred miles long (on average about 400 miles (644 kms)) whereas the deepest part of the ocean is only seven miles deep. By using the calculation from earlier, we know that the water depth is clearly less than 1/20 of the wavelength.

(10) *Inshore waves* are slightly different to shallow water and deep-water waves. The length of these waves is less than the depth of the water they enter, which decreases the velocity of the waves. This results in a shortened wavelength and an increase in wave height. As the imbalance of the wave increases, the wave collapses and breaks into a backwash.

(11) *Internal waves* are one of the largest types of waves encountered at sea, yet they are also one of the least noticeable. This is due to their formation in the internal layers of the water column. Ocean water comprises different layers. Towards the bottom of the column is colder saline water. As the depth rises, the temperature increases, and the salinity decreases. When the interface between high saline water and low saline water is disturbed − perhaps by tidal movements − internal waves are generated. Similar in shape and structure to surface waves, internal waves travel long distances, whilst accruing energy and velocity. When they reach a landmass, the resultant destruction is very similar to that which occurs with tsunami waves. It is quite common for internal waves in the Luzon Strait, South China Sea, to reach 200 metres (or approximately 660 feet) in height.

(12) *Kelvin waves* are named after their discoverer, Sir William Thompson, later Lord Kelvin. Kelvin waves are typically found in the Pacific Ocean. They are large-scale waves formed by a lack of wind flow. Unlike normal waves, Kelvin waves are a special type of gravitational wave that are influenced by the Earth's rotation and are trapped at the equator or along vertical lateral boundaries such as coastlines and underwater mountain ranges.

(13) *Progressive waves.* For a progressive wave, the amplitude is equal to all other points, resulting in a net energy flow. In other words, they are a form of wave in which the ratio of an instantaneous value at one point is constant compared to all other points. There are three types of progressive waves: longitudinal, transverse, and orbital.

(14) *Capillary waves* closely resemble ripples on the water's surface. The restoring force involved is capillarity, which is the binding force that holds water molecules together. The particularly wavy structure of capillary waves is caused by light breezes and calm winds that blow at speeds of around 3–4 metres (9.84–13.12 ft) per second and at a reference height of 10 metres (16 ft) from the water's surface. Typical capillary wavelengths are less than 1.5 cm (0.59 inches) and have a duration of less than 0.1 seconds.

(15) *Refracted waves*. These waves travel in shallow water are most noticeable when they approach the shore. The shallowness of the water decreases the power of the wave, causing it to curve round. Refracted waves are most commonly found in bays and around headlands.

(16) *Seiche waves*, pronounced 'saysh'. These are standing waves that form in confined or partially confined waters. When water sloshes back and forth in a confined space, it forms seiche waves. On a small scale, this might be in a swimming pool, a bucket or even a glass. On a larger scale, seiche waves are found in lagoons and large lakes. Seiche waves are generated when either rapid changes in the atmospheric pressure or strong winds force the water towards one end of the confined space, causing it to 'pile up'. When the external force stops exerting pressure on the water, the 'piled-up' water – possessing potential energy – rebounds back towards the opposite end of the enclosed water body. This periodic oscillation of water, without anything to offer resistance, can continue unabated for long periods of time (hours or even days). It is not uncommon for seiche waves to be mistaken for tides. This is because the period of time it takes for the wave to change from a trough to a crest can be anything up to 7–8 hours, which is roughly comparable to most tides. Although the causal factors may be the same for seiche waves and tsunamis, seiches are fundamentally different. Seiches are essentially standing waves with long periods of oscillation and occur almost exclusively in enclosed bodies of water, whereas tsunamis are progressive waves occurring in free bodies of water.

In this chapter we have covered some of the basic principles of wind, current and water depth and how these contribute to the formation of different wave types. Waves are a natural phenomenon. They are a source of sporting fun, but they are also a source of danger and destruction. For the ship's navigator, wave patterns and changes in wave activity can provide crucial clues to the weather and any potential changes in sea conditions. In the next chapter, we will turn our attention to the role and function of navigation marks and buoys.

2   Navigation marks and buoys

Navigation marks and buoys are an indispensable part of safe maritime navigation. Just as road users must follow road signs and traffic lights, ships' navigators must follow a uniform system of marks and buoyage. The development of a uniform system of buoyage is one of the greatest improvements in global shipping. To establish a system that could be followed anywhere around the world, the International Association of Marine Aids to Navigation and Lighthouse Authorities (IALA) developed a set of rules called the *IALA Buoyage System for Mariners*. To maintain worldwide uniformity, the IALA divided the world into two regions: *Region A* and *Region B*. Region A encompasses Europe, Australia, New Zealand, Africa, the Gulf States, and some countries in Asia. Region B incorporates North, Central and South America, Japan, Korea, and the Philippines. To enhance the safety of the system, the IALA assigned colours to the port and starboard sides of navigable channels. These sides would then be marked using coloured buoys called lateral marks. This system is called the *Lateral System*. In Region A, the colour red is used to designate the port side of the channel and the colour green is used to designate the starboard side of the channel. In Region B, the colours are reversed. The IALA Buoyage System provides six types of marks that navigators must observe: (1) lateral marks; (2) cardinal marks; (3) isolated danger marks; (4) safe water marks; (5) special marks; and (6) emergency wreck marking buoys. Importantly, regional colour variations do not apply to cardinal, isolated danger marks, safe water marks, or special marks.

Lateral marks

Lateral marks were designed to help indicate which side of a waterway a ship must follow. The system dictates that port marks should be kept to the ship's left side and starboard marks to the ship's right side. That said, when a ship is sailing downstream, the position of the marks will change accordingly, i.e., port marks will appear on the right side and starboard marks will appear on the left side. When a channel splits to form more than one potential direction of travel, a modified lateral mark is used to indicate the "preferred channel". A preferred channel is indicated by red and green horizontal bands on the lateral mark. Sometimes, lateral marks are sequentially numbered. This shows the navigator the conventional direction of buoyage. Every lateral mark is identifiable by their colour, shape, top marks, light and light rhythm. In Table 2.1, we can clearly see the distinction between Region A and Region B lateral marks:

DOI: 10.1201/9781003291534-2

Table 2.1 Lateral marks, regions A and B.

Lateral Marks: Region A

	Port Side Marks	Starboard Marks
Colour	Red	Green
Buoy Shape	Cylindrical (can); pillar; spar	Conical; pillar; spar
Top Mark	Single red cylinder (can)	Single green cone pointing upwards
Light Colour	Red	Green
Light Rhythm	Any apart from composite group flash (2 + 1)	Any apart from composite group flash (2 + 1)

Preferred Channel Lateral Marks: Region A

	Preferred Channel to Starboard	Preferred Channel to Port
Colour	Red; green; red horizontal stripes	Green; red; green horizontal stripes
Buoy Shape	Cylindrical (can); pillar; spar	Conical; pillar; spar
Top Mark	Single red cylinder (can)	Single green cone pointing upwards
Light Colour	Red	Green
Light Rhythm	Composite group flash (2 + 1)	Composite group flash (2 + 1)

Lateral Marks: Region B

	Port Side Marks	Starboard Marks
Colour	Green	Red
Buoy Shape	Conical; pillar; spar	Conical; pillar; spar
Top Mark	Single green cylinder (can)	Single red cone pointing upwards
Light Colour	Green	Red
Light Rhythm	Any apart from composite group flash (2 + 1)	Any apart from composite group flash (2 + 1)

Preferred Channel Lateral Marks: Region B

	Preferred Channel to Starboard	Preferred Channel to Port
Colour	Green; red; green horizontal stripes	Red; green; red horizontal stripes
Buoy Shape	Cylindrical (can); pillar; spar	Conical; pillar; spar
Top Mark	Single red cylinder (can)	Single green cone pointing upwards
Light Colour	Green	Red
Light Rhythm	Composite group flash (2 + 1)	Composite group flash (2 + 1)

Cardinal marks

Cardinal marks are used in conjunction with the ship's compass to indicate where the mariner may find the best navigable waters. They take their name from the quadrant in which they are placed. Cardinal marks have the same colour and same shape, irrespective of whether they are in Region A or Region B. As per the points of a compass, there are four cardinal marks: *North, South, East*, and *West*. Each mark is distinguishable from one another by their top marks, buoy colour and the rhythm of their light. When a cardinal mark is seen, it indicates the water on the named side is clear and navigable. In other words, if travelling on an easterly course and you see a North cardinal mark ahead, this indicates the direction of passage should be north of the cardinal mark, requiring a change in course to port. Cardinal marks are also used to indicate features or hazards in the channel. These might include bends, junctions, branches, or the end of a shoal. Though this might seem complicated to start with, it is quite simple to commit cardinal marks to memory if we remember that North and South cardinal marks follow a North and South direction; North cardinal marks have an upward-pointing top mark whereas South cardinal marks have a downward-pointing top mark. East cardinal marks have an egg-shaped top mark, not dissimilar to an Easter egg. West cardinal marks have a tapered top mark, which may be associated with a woman's waist. The rhythm of light can be related to the face of a clock. All cardinal marks exhibit a white light.

Table 2.2 Cardinal marks (North, East, South and West)

Cardinal Marks: North and East

	North Cardinal Marks	East Cardinal Marks
Colour	Black above yellow	Black with single horizontal yellow band
Buoy Shape	Pillar; spar	Pillar; spar
Top Mark	Two black cones pointing upwards	Two black cones; base to base
Light Colour	White	White
Light Rhythm	VQ or Q (uninterrupted)	VQ(3) or Q(3)

Cardinal Marks: South and West

	South Cardinal Marks	West Cardinal Marks
Colour	Yellow above black	Yellow with single horizontal black band
Buoy Shape	Pillar; spar	Pillar; spar
Top Mark	Two black cones pointing downwards	Two black cones; pointing towards each other
Light Colour	White	White
Light Rhythm	VQ(6) + long flash or Q(6) + long flash	VQ(9) or Q(9)

Isolated danger marks

As the name suggests, these buoys are used to mark dangers and hazards to shipping. Isolated danger marks are erected or moored directly above the location of the danger to alert navigators of the hazard ahead. An isolated danger mark also indicates that there is safe and navigable water around the mark. These marks are distinguishable from all other marks by their top marks, which consists of two black spheres one above the other. They also have different colours to all other marks; black with one or more red horizontal bands. The rhythm of the light is unique following a group flashing two.

Table 2.3 Isolated danger marks

Colour	Black with one or more red horizontal bands
Buoy Shape	Pillar; spar
Top Mark	Two black spheres; vertically aligned
Light Colour	White
Light Rhythm	Group flashing 2

Safe water marks

Unlike the other marks which use horizontal stripes or bands, safe water marks are the only buoys to use vertical stripes. The safe water mark is used to indicate several safe water areas such as the mouth of marked channels, demarcating the boundary between the open sea and confined waters, the point of entry into a harbour or port, and the best point of passage under fixed overhead obstructions such as bridges. Safe water marks may also be positioned in a line to mark out the safe navigable route through shallow areas.

Table 2.4 Safe water marks

Colour	Red and white vertical stripes
Buoy Shape	Pillar; spar; spherical
Top Mark	Single red sphere
Light Colour	White
Light Rhythm	Isophase, occulting, 1LF every 10 seconds or Morse Code 'A'

Special marks

Special marks are used to indicate areas with special features and, as such, do not play any major role in facilitating safe navigation. Special marks may indicate spoil grounds, military exercise areas, recreational zones, anchorage boundaries, undersea cables and pipelines, dead ends, mooring areas, protected areas, marine and aquaculture farms, oil wells, and Ocean Data Acquisition System (ODAS) modules.[1] The nature of the special mark can be easily ascertained by looking at the appropriate chart or Sailing Directions. Special marks are identifiable by their yellow colour and top mark, which consists of a single yellow X-shaped cross.

Table 2.5 Special marks

Colour	Yellow
Buoy Shape	Optional but must not conflict with those used for lateral or safe water marks
Top Mark	Single yellow X-shaped cross
Light Colour	White
Light Rhythm	Any rhythm not used for white light

Emergency wreck marking buoy

These marks came into existence much later than the other five marks discussed above. In 2002 the Norwegian car carrier MV *Tricolor* collided with another ship in the English Channel before capsizing and partially submerging. Over the next couple of months, a further two ships collided with the MV *Tricolor*. These incidents prompted the IALA to incorporate emergency wreck marking buoys to the IALA Buoyage System as the IALA recognised there was a need to mark out new shipping hazards that would not appear on current charts or in extant navigational publications. These buoys are designed to be placed as close to the wreck as possible and, unlike other buoys, are designed to provide a highly conspicuous visual and radio aid to navigation.

Table 2.6 Emergency wreck marking buoy

Colour	Yellow and blue stripes – minimum four; maximum eight
Buoy Shape	Pillar; spar
Top Mark	Upright yellow cross (+)
Light Colour	Yellow
Light Rhythm	Any light rhythm not used for white light (B 1.0S + 0.5S + Y 1.0S + 0.5S)

Traffic separation schemes

As well as indicating dangers and hazards, and showing the best direction of travel, marks and buoys are extensively used in Traffic Separation Schemes (TSS). As the incident with the MV *Tricolor* demonstrates, separating ships in opposing directions of travel in confined water channels is critical to ensuring safe navigation. This is particularly evident in extremely congested waters such as those off the coast of China and Singapore. According to the International Maritime Organisation (IMO), there are over 200 TSS in operation across the globe. TSS are indicated by purple dashed lines on admiralty charts. Although they exist on charts, in reality, TSS are imaginary divisions that are designed to separate ships travelling in opposite directions. Compliance with TSS demarcations is compulsory as per Rule 10 of the COLREGS.[2] Despite compliance being mandatory, incidents and collisions do frequently occur within channels designated with TSS divisions. There are several reasons why these incidents happen, which we will briefly discuss here.

(1) *Assuming that navigating within a traffic lane provides right of way over other traffic.* This is a common cause for incidents occurring within TSS though in fact rule 10 of the COLREGS states "this rule applies to traffic separation scheme adopted by the Organisation [IMO] and does not relieve any vessel of her obligation under any other rule." In other words, rule 10 stipulates navigators must comply with the COLREGS at all times irrespective of whether or not the ship is in a TSS. Being in a TSS does not relieve the navigator of their duty to comply with the COLREGS. To put this into a practical example, a vessel engaged in fishing must not impede the passage of a ship within a TSS, nor may a small craft (any vessel less than 20 metres (66 ft) in length of a sailing vessel) impede a ship's passage. The only exception to rule 10 is where the risk of collision is so great that evasive action is required, as per rule 18 of the COLREGS.

(2) *Crossing at incorrect angles.* Rule 10c of the COLREGS states that "a vessel shall, so far as practicable, avoid crossing traffic lanes but if obliged to do so shall cross on a heading as nearly as practicable at right angles to the general direction of traffic flow." This means that any vessel crossing ahead of another vessel must do so at a heading which is as close to right angles to the general flow of traffic as possible. The reason for this is twofold: first, heading at right angles results in a shorter crossing time; and second, the intent of the ship crossing ahead at right angles is more visually obvious than a ship crossing at a lesser angle.

(3) *Laying course on top of the purple arrows and navigating exactly in the centre of the lane.* Rule 10b of the COLREGS states that "a vessel using a traffic separation scheme shall:

   (a) Proceed in the appropriate traffic lane in the general direction of the traffic flow for that lane.
   (b) So far as practicable keep clear of a traffic separation line or separation zone.
   (c) Normally join or leave a traffic lane at the termination of the lane, but when joining or leaving from either side shall do so at as small an angle to the general direction of traffic flow as practicable."

Charts often display arrows pointing in the general direction of the traffic flow. These are merely indicative only. Subsequently, navigators must lay waypoints along the intended course, bearing in mind that alterations may be required to avoid collisions or navigational hazards. For fast ships, this means leaving sufficient sea room either side of the vessel to allow for safe overtaking. Slow steaming or super-slow steaming ships must never navigate in the middle of the lane as this presents a hindrance to other traffic and inhibits safe overtaking by faster vessels.

(4) *Ignorance of vessel position reporting requirements.* Vessel position reporting is an important part of safe and considerate navigation. Very often, traffic separation schemes will have mandatory position reporting requirements, and these are especially so when part of an IMO or local routing scheme. In accordance with SOLAS[3] V/10, a ship must report their position when mandated to do so unless there is a compelling reason not to, in which instance a record must be made in the ship's log. Furthermore, SOLAS V/11 requires the ship's master to comply with the reporting requirements and to report all relevant information to the competent authorities. Some reporting requirements follow specific formats such as GIBREP, COPREP, FINREP, CALDOVREP, and WETREP.[4] It is quite common to hear Vessel Traffic Services (VTS) admonishing non-compliant vessels over VHF for breaching the mandatory reporting requirements.

(5) *Insufficient appraisal of the passage plan.* No matter how many times a navigator has passed through a TSS, they must carry out a full appraisal of the passage plan each and every time. This is important as major changes may have occurred since the last passage through the TSS. New wrecks may be marked, oil rigs or offshore installations may have relocated, or survey activity might be undertaken. Importantly, every TSS has its own unique hazards. Passing through the Singapore and the Malacca Strait, for example, is very different to navigating the English Channel or the Gulf of Suez. Furthermore, the movement of traffic must be considered and regularly accounted for. Each ship will have a different intent. A vessel may be approaching or departing port, in which case pilot transfer may be in progress. A ship ahead may be planning to execute a sharp turn to overtake a slower vessel in their path. A ship may slow down for replenishment of stores or crew change whilst remaining underway. All these potential hazards need to be considered and frequently reappraised.

(6) *Inadequate planning, execution, and monitoring.* It is the responsibility of every ship's navigator to ensure they undertake their duties with utmost care and attention. This includes planning, executing, and monitoring the ship's manoeuvres based on the vessel's current characteristics and status. Some of the key points to consider is the ship's draught; the nature of the cargo being carried onboard; any critical areas; and the effect of tidal streams and eddies on vessel speed and handling.[5] Once the frequency and method of position fixing has been decided, it is important this is strictly adhered to. Often due to the proximity of coastal landmarks, parallel indexing may be the easiest and quickest method of continuous position marking.

(7) *Lack of familiarity with the ship's Safety Management System (SMS).* Many ship operators provide guidelines pertaining to navigation and bridge watchkeeping in their SMS. These may include generalised guidelines for safe practices to be followed in TSS or very specific and detailed instructions for a particular

TSS. In the latter instance, this will typically include details on the bridge and engine room manning levels. Some companies may also stipulate the addition of extra officers on the bridge, and even the minimum rank and duties of the *Officer of the Watch* (OOW). There will also be guidance on machinery status – i.e., whether the main engine should be on standby, whether the engine control room should be manned, and whether or not auxiliary engine machinery should be running.

(8)   *Complacency.* Whilst TSS have undoubtedly succeeded in reducing collisions by avoiding head-on meetings and reducing the relative speed of approach for ships in the same traffic lane, accidents and incidents still happen with alarming frequency. Therefore, the prudent navigator must always exercise extra vigilance and caution when transiting areas with traffic separation schemes in place.

Although the increased presence of navigational markers and buoys has done much to improve the job of the modern mariner, there sadly remains the issue of human imperfection. Complacency, laziness, poor training, and lack of oversight; these are all contributing factors to maritime accidents; all of which are perfectly avoidable. It is not just human failings though that can lead to the loss of life and marine devastation. Fog, heavy rains, snowfall, sandstorms or even smoke from fires can result in restricted visibility. The main engines, steering gear, auxiliary engines, anchoring and mooring equipment; and virtually any other type of machinery carried onboard – no matter how well maintained – can malfunction at the worst possible time. To avoid or minimise the consequences of these hazards, it is critical contingency measures are in place and followed. The creation of TSS has made navigation much safer, but it has not removed the hazards and dangers of sailing completely.

Ship navigators are well advised to have a complete understanding of the COLREGS, as well as an appreciation of real-life scenarios to help guide them as they navigate through difficult and challenging waters, which is a topic we will discuss in greater depth later in this book.

Figure 2.1 Port Hand Buoy on the Tom River, Tomsk, Russia.

Figure 2.2 Starboard Hand Mark, Kyle of Lochalsh, Scotland.

Figure 2.3 North Cardinal Mark, Thames Estuary, England.

Figure 2.4 East Cardinal Mark, Bremerhaven, Germany.

Figure 2.5 South Cardinal Mark, St Mawes, England.

Figure 2.6 West Cardinal Mark, Gigalum Rocks, Scotland.

Figure 2.7 Safe Water Mark, Limfjord, Denmark.

Figure 2.8 Isolated Danger Mark buoy, Hvar, Croatia.

Figure 2.9 Special Mark, Queensland, Australia.

Notes

1 These are scientific modules which collect data such as windspeed, air pressure, water salinity and water temperature.
2 International Regulations for Preventing Collisions at Sea, 1972.
3 International Convention for the Safety of Life at Sea, 1974.
4 Ship reporting requirement for the "Strait of Gibraltar"; "Off the Coast of Portugal"; "Off the Coast of Finland"; Calais Dover Strait (English Channel); Western European Tanker Reporting System.
5 Depending on the ship's draught, the navigator may need to consider planning for or altering course from the regular route to a deep water route; restrictions may apply regarding the distance a vessel must maintain from a coastal states shore if carrying hazardous cargo in bulk or IMDG cargoes; finally, it might be more appropriate for a vessel to transit a specific area during daylight hours owing to the vessel's condition, restricted sea room, or the presence of heavy traffic.

3   Navigation systems and equipment

Ships, by necessity and design, carry a vast selection of navigational equipment and systems. Some of this equipment, such as the sextant, are almost as old as seafaring itself; others are a truly modern invention. In this chapter, we will briefly look at the role and function of some of the main navigational equipment and systems used by contemporary ship navigators.

Navigation equipment and aids

Gyro compass

The gyro compass is a form of gyroscope used widely on ships, employing an electrically powered, fast-spinning gyroscope wheel. Unlike the magnetic compass, gyroscopic compasses are not affected by external magnetic fields. This means they are ideal for locating true north; from true north, an accurate and stable direction can be plotted. The gyro compass has become one of the most indispensable instruments carried on board modern merchant and naval ships for its ability to detect the direction of true north and not magnetic north. The gyro compass consists of several components, including the (1) Master compass: this discovers and maintains the true north reading with the help of a gyroscope; (2) Repeater compass: this receives and indicates the true direction transmitted electronically from the master compass; (3) Course reader: this makes a continuous record of the manoeuvring on a moving strip of paper; (4) Control panel: governs the electrical operation of the system and ascertains the running condition by means of a meter; (5) Voltage regulator: maintains a constant supply of voltage from the ship's motor-generator to the compass; (6) Alarm unit: indicates any failures of the ship's supply; (7) Amplifier panel: controls the follow-up system; (8) Motor generator: converts the ship's DC supply to AC and energises the compass equipment. The gyro compass is linked to the repeater compass via a transmission system. The fast-spinning rotor weighs from as little as 0.56 kgs (1.25 lbs) to as much as 24.94 kgs (55 lbs). It is driven thousands of revolutions per minute (rpm) by another electric motor. The most critical part of the compass though is the spinning wheel, which is also known as the gyroscope.

DOI: 10.1201/9781003291534-3

Figure 3.1 Gyroscope compass, HMCS *Algonquin.*

External magnetic fields which deflect normal compasses have no effect on gyroscopic compasses. When a ship alters course, the independently driven framework called the "phantom" moves with the ship, yet the rotor system continues to point northwards. This lack of alignment enables the compass to send a signal to the driver motor, which moves the phantom in step with the rotor system again in a path where the phantom may have crossed only a fraction of a degree to several degrees of the compass circle.

As soon as the phantom and the rotor system are realigned, the phantom sends electrical pulses to the repeater compass for each degree of difference. The gyroscope in the gyroscope compass is mounted in such a way that it can move freely about three mutually perpendicular axes and is controlled so as to enable its axis to spin settled parallel with the true meridian, influenced by the Earth's rotation and gravity. The manner in which the gyroscope compass works is based on two fundamental characteristics:

(1) gyroscopic inertia; and
(2) precession.

Gyroscopic inertia is the tendency of any revolving body to uphold its plane or rotation. Precession is a property that causes the gyroscope to move when a couple is applied. But, instead of moving in the direction of the couple, the gyroscope moves at right angles to the axis of the applied couple and the spinning wheel. These two properties – when combined with the Earth's two natural forces of rotation, and gravity – allows the gyro compass to seek true north. Once settled on the true meridian, the rotor will indefinitely remain constant for as long as there is an electrical current. Despite gyro compasses becoming preeminent on most, if not all, modern ships, and watercraft, they are not infallible, and errors may occur. For example, on northerly courses it has been found the gyro compass is slightly deflected to the west of true meridian whereas on southerly courses, it is deflected to the east. Most ships use the Global Positioning System (GPS) or other navigational aids to feed data to the gyro compass to compensate for the error. An orthogonal triad of fibre optic design and ring laser gyroscopes, which apply the principles of optical path difference to determine the rate of rotation, instead of depending on mechanical parts, have shown it is possible to eliminate these flaws and gain a detection of true north.

Magnetic compass

The magnetic compass works in conjunction with the magnetic field of the Earth. It is used to plot a planned course for the ship. The magnetic compass is usually fitted at the centre line of the ship, high up on the monkey island. A transmitter sends readings direct to the bridge and these are displayed on the bridge panel. The importance of the magnetic compass cannot be overstated. In fact, SOLAS Chapter 5, regulation 19, states "all ships irrespective of size shall have a properly adjusted standard magnetic compass or other means, independent of any power supply to determine the ship's heading and display the reading at the main steering position".

Binnacle

Owing to the exposure of the magnetic compass to the weather and harsh sea conditions, it is necessary to protect the compass from the elements. The navigation instrument that serves this purpose is the binnacle. As well as protecting the compass from the weather, the binnacle also stops the compasses from toppling over due to the constant rolling and pitching of the vessel. Binnacles have been a feature on ships since at least the 1700s. Over the years though, the binnacle has undergone technological development as ships have gone from wood to iron to steel. The binnacle is a cylindrical container made from non-ferrous material, though early binnacles were made exclusively from hardwood. The instrument consists of a round stand upon which a plinth is mounted. Separate constructions accommodate the compass and other navigational equipment. The compass bowl is housed in the top part of the binnacle, and in the middle part, accessible through a door, the corrector magnets extend to the projector and out towards the fore. The binnacle is always positioned right before the helmsman's line of vision. The compass and all other navigation equipment are held within the binnacle, making it easy for the helmsman to observe the course and steer the vessel accordingly. For this reason, binnacles are traditionally made to average waist height.

Within the binnacle are several magnetic compass correctors. These are the:

(1) Flinders Bar
(2) quadrant spheres
(3) heeling magnet
(4) athwartships magnet
(5) fore and aft magnetics.

These corrector magnets help to negate any disturbing effects of magnetism caused by the hull of the ship. The exact specifications of the corrector magnetics are decided by an approved 'compass adjuster' whose services are usually called upon when the ship is in port. The access door to these corrector magnets must always be kept locked unless when access is needed for adjustment. The Flinders Bar is a soft iron corrector which is placed vertically in the forward or aft part of the binnacle depending on the extension of the superstructure area abaft of or forward of the binnacle. The length of the Flinders Bar may change during any authorised adjustment of the compass. The Quadrantal Correctors are two soft iron spheres fitted in brackets on either side of the binnacle. Again, the distance between the quadrantal correctors can be altered during an adjustment of the compass. A helmet is fitted on top of the binnacle and is made of non-ferrous material. This is in place to protect the compass bowl from the elements and harsh sea conditions.

The equipment held within the binnacle is very sensitive and critical to the ship's safe navigation. For this reason, it is imperative the binnacle is properly maintained. To keep the binnacle and the equipment held inside in prime condition:

(1) always keep the access door locked
(2) if made of wood, the binnacle should be varnished and never painted
(3) the quadrantal corrector must be painted from time to time to prevent rust from settling
(4) the brass parts should be polished regularly
(5) any materials with magnetic properties must be kept away from the compass
(6) the protective helmet should always be kept in place
(7) any adjustments must be authorised and carried out by an approved compass adjuster
(8) any bubbles in the compass bowl must be removed.

To remove bubbles from the compass ball:

(1) tilt the bowl until the filler hole can be seen
(2) unscrew the locking screw
(3) top up with ethyl alcohol – if ethyl alcohol is not available, use distilled water instead
(4) screw the locking screw back in place
(5) return the bowl to its original position.

Connected to the binnacle is the compass projector. As we noted above, the magnetic compass is located on the monkey island, whereas the compass course needs to be read from the bridge. The compass projector enables the helmsman to see the compass by way of several reflections. These enable the helmsman to be at the wheel in the bridge, able to see the compass card reading and lubber line to steer the ship in the correct course.

Navigational Telex (NAVTEX)

It is of utmost importance that the ship's navigator ensures the safety of their vessel and her crew. Accidents can happen even to the most cautious and prudent navigator. Starting from the beginning of the passage planning, the navigator needs up-to-date information that will potentially impact on the ship's progress. The most important type of information for navigators that is information related to safety, including Maritime Safety Information (MSI) bulletins. MSI includes navigational and meteorological warnings, meteorological forecasts, warnings about dangers to navigation, warnings of missing vessels and a variety of other urgent messages pertaining to the safety of the vessel and her crew. Constant monitoring to pick up wanted information among a vast volume of messages is not an easy or even a practical task with a limited radio system. In answer to this, the NAVTEX system provides navigators with up-to-the-minute information automatically. NAVTEX, an acronym for navigational telex (navigational text messages), is a device found on board vessels to provide short range Maritime Safety Information in coastal waters. It is used by ships of all types and sizes. The area covered by NAVTEX can extend as far as 400 nautical miles from the broadcast station. A NAVTEX receiver onboard prints out navigational and meteorological warnings and forecasts in addition to urgent Marine Safety Information messages. It forms a vital component of the Global Maritime Distress and Safety System (GMDSS). NAVTEX uses radio telex or Narrow Band Direct Printing (NBDP) for receiving the automatic broadcast of information.

In essence, NAVTEX is a form of extra insurance. It is a very convenient way of monitoring navigational warnings, meteorological warnings, search and rescue information and other data for ships sailing within (230 miles - 460 miles; 370 kilometres - 740 kilometres) off the coast. It thus provides navigational and weather-related information in real time. As the NAVTEX receiver receives messages without user input, it is extremely user-friendly. The navigator or OOW is not required to monitor the NAVTEX regularly or to be physically present at a fixed time.

There is also no requirement for retuning the receiver. This not only saves time but also stops an officer from being distracted on the bridge. With the information received from the NAVTEX receiver, the passage plan can be amended as required. The OOW can attend to any distress warning in the vicinity, and they are appraised of the expected weather and can plan accordingly. Thus, a NAVTEX forms an integral part of the bridge navigational equipment.

How NAVTEX works

The NAVTEX system works on a frequency of 518 kHz in the medium frequency band. 490 kHz frequency is also used by some countries for broadcasts in their national languages, which is known as national NAVTEX. Where medium frequency reception is difficult, transmissions are made on 4209.5 kHz. The default setting for NAVTEX is 518 kHz. The world is divided into 21 areas known as navareas (including five areas introduced for the Arctic region). Each navarea serves the purpose of distributing MSI and related information. Every navarea has multiple NAVTEX stations. All NAVTEX receivers are programmable to enable the navigating officer to ensure that only messages from selected NAVTEX stations are displayed or printed.

The SELECTING STATION menu under the Menu option on the NAVTEX receiver allows the officer to select the desired stations they want to receive automatically or manually. On automatic selection, the NAVTEX receives continuous MSI for the area the ship happens to be in and without any user involvement.

If a ship's position data are fed from any navigating equipment such as GPS, the NAVTEX will automatically decide which navarea the ship is navigating in and thus selects the appropriate NAVTEX station. In manual mode, the navigating officer can select what stations they want to receive. A list of NAVTEX stations can be found in the Admiralty List of Radio Signals Vol. 3 Pt 1 and in the List of Coast Stations and Special Service Stations (List IV) for reference.

Format of NAVTEX messages

NAVTEX messages are separated into different types depending on the purpose and content of the broadcast. Accordingly:

A = Navigational Warning
B = Meteorological Warning
C = Ice report
D = Search and Rescue Information; piracy and armed robbery
E = Meteorological forecast
F = Pilot messages
G = AIS messages (formerly Decca messages)
H = Loran C messages
I = Omega messages
J = Satnav messages (GPS or GLONASS)
K = Other electronic navigational aid system messages
L = Navigational warnings (additional)
M to U = Reserved
V = Notices to fisherman
W to Y = Reserved
Z = No messages on hand

The NAVTEX receiver can be set to ignore certain types of messages; however, messages A,B,D and L because of their importance cannot be declined by navigating officers. Audible alarms can also be generated when message type A,B,D or L are received. It is only possible to reset these alarms manually. When programming the type of messages to receive, it is prudent to ensure that only those which are required and necessary are programmed. Not doing so will waste an inordinate amount of paper but also risks losing important information in the noise. When a NAVTEX message is received, it appears in a set format, for example:

ZCZC b1 b2 b3 b4   MAIN MESSAGE   NNNN

We can interpret the message accordingly: ZCZC: This is the start code. It indicates the beginning of the message; B1: This character represents the Station ID; B2:

This character is called the Subject Indicator and is used to represent the type of message (A to Z). The characters B1 and B2 are used by the NAVTEX receiver to reject messages from stations concerning subjects of no interest to the officer; B3 and B4: B3 and B4 is a 2-digit serial number for each message. The characters B3 and B4 are used by receivers to keep an already received message from being repeated; NNNN: This indicates the end of the message. Below is an example of a typical NAVTEX message:

ZCZC OA20
WZ 1593
Scotland, West Coast
The North cardinal light buoy 58.01.2 N 005.27.1 W
has been permanently withdrawn.
Cancel WZ 1562
NNNN

Every NAVTEX message has information within a message header. In the above message: "O" indicates a broadcast from the NAVTEX station, in this example, Port Patrick radio; "A" indicates a Navigational warning category message; '20' indicates the navigational warning message priority sequence.

Practical advice

Every officer should make sure that there are always sufficient rolls of NAVTEX paper available onboard. It is important to check that there is paper in the receiver so that no important messages are missed. It is advisable to always leave the NAVTEX on to avoid the chance of losing vital information that might impact the vessel during its voyage. Also, always ensure the operating manual is available on the bridge. A plastic copy of the navareas/metareas where the vessel is likely to sail, showing the NAVTEX stations, their coverage ranges, and their respective time schedules, should be made available next to the NAVTEX equipment. Routine tests should be carried out to check the performance of the equipment. Extra care should be taken not to confuse the programming of B1 characters (station designators) with those of B2 characters (type of messages).

Automatic Radar Plotting Aid (ARPA)

The Automatic Radar Plotting Aid (ARPA) is a system which displays the position of the ship and all other vessels within a set proximity. The ARPA then selects the most appropriate course for the ship to avoid collision. The system constantly monitors the ship's surroundings, automatically acquiring and deleting targets as they come into the ARPA's range. ARPA has the added benefit of not only tracking other vessels, but is also capable of tracking smaller craft, stationary objects, and floating debris. ARPA represents these targets as vectors on the display screen and constantly updates their parameters with each turn of the antenna, calculating their nearest points of approach to the ship.

Automatic tracking aid

Just like ARPA, the automatic tracking aid displays the information of tracked targets in graphic and numeric format. This generates a planned layout for a safer and collision-free course. Usually, when an echo measures 800 metres (2,624 ft) or more in circumference, the system interprets these targets as 'landmass' and they are therefore not tracked. Any echoes less than 800 metres (2,624 ft) are considered targets and are tracked accordingly.

Speed and distance log device

This piece of bridge equipment is used to measure the speed and distance travelled by the ship from a set point. By calculating the speed and distance travelled, it is possible to determine the ship's estimated time of arrival (ETA).

Global Positioning System (GPS) receiver

The Global Positioning System (GPS) receiver is a display system used to show the ship's location with the help of global positioning satellites located in the Earth's atmosphere. With the record of the ship's position, speed, and course, the navigator can easily and accurately calculate the time taken to cover the distance between "two marked positions".

Echo sounder

The echo sounder is used to measure the depth of water below the ship's keel. The device works on the principle of transmitting soundwaves and an audio pulse which then bounces off any reflecting layers beneath the ship. This returns to the sounder as an echo. The depth is then calculated according to the time taken for the soundwave to transmit, bounce and return to the sounder.

Long-Range Identification and Tracking System (LRIT)

LRIT is an international tracking and identification system incorporated by the IMO into the SOLAS Convention to facilitate the tracking of all ships over 300 gross tonnes on international voyages.

Rudder angle indicator

The rudder angle indicator, as the name suggests, provides the helmsman with the angle of the rudder. The display is provided on the navigation bridge equipment console so that the ship's navigator can control the rate of turn and the rudder angle of the ship. The indication is also provided in the bridge wing and in the engine control room.

Rate of turn indicator

The rate of turn indicator is a tool which shows how fast the ship is turning at a steady rate. This is particularly useful during pilotage and manoeuvring operations. The rate of turn is shown as each degree turned in 60-second intervals.

Voyage Data Recorder (VDR)

The Voyage Data Recorder (or VDR) is a crucial instrument among the ship's navigation equipment list. It is installed on ships to continuously record vital information related to the operation of the vessel. It contains a voice recording system which covers a rolling 12-hour period. This recording may be recovered and used for investigation purposes. In essence, the VDR is identical in its purpose to the eponymous "black box" found on aeroplanes.

Transmitting Heading Device

The Transmitting Heading Device (or THD) is an electronic device which is used to display the information of the vessel's true heading. The THD compliance information is provided in Chapter 5 of the SOLAS Convention.

Sound reception system

This acoustic system is required, as per the COLREGS, for ships with a fully enclosed bridge. It enables the ship's navigator, who is located inside the sealed cabin, to listen to sound signals (such as the fog or ship's horn) from other ships in the vicinity.

Ship's whistle

The ship's whistle, also known as the ship's horn, is located on the bridge, and is provided in two forms: one is driven by air and the other is electrically operated. Amongst the other instruments used in difficult navigation such as bad weather, fog, poor visibility, and high traffic, the ship's whistle helps to alert other vessels nearby. In emergency situations, the ship's whistle is used to notify the ship's crew and to alert other vessels nearby to render assistance.

Daylight signalling lamp

These are light-signalling devices used for emergency signalling during daylight hours and may also be used in darkness.

Pilot card

The pilot card is an informative booklet provided for the use of the ship's pilot. It consists of the dimensions, draught, turning circle, manoeuvring parameters, propulsion equipment and other navigation tools and instrument lists of the vessel necessary for safe navigation.

Fo'c'sle bell

The fo'c'sle (or forecastle bell) is used to mark the presence of the ship in fog or bad weather. It is also sounded as an alarm in the event of an emergency, together with the ship's whistle or horn.

Manoeuvring booklet

In this booklet, the performance parameters of the propulsion plant and the ship's manoeuvring capabilities in different weathers and situations are recorded for quick reference. This typically includes the following: (1) the ship's general description; (2) the ship's manoeuvring characteristics in deep water; (3) the ship's stopping and speed control characteristics in deep water; (4) the ship's manoeuvring characteristics in shallow water; (5) the ship's manoeuvring characteristics in wind; (6) the ship's manoeuvring characteristics at low speed; and (7) any additional relevant information.

Record of navigational activities

All navigational activities which are performed by the ship's officers and crew using different navigation equipment on the bridge must be recorded and kept on board for reference. This is mandatory and is the most important part of the ship's logbook.

Record of maintenance of navigational equipment

A hard copy of all the ship's navigation systems and equipment list must be recorded and present on board the ship for inspection by port and regulatory authorities. It must be signed by the ship's master and the duty officers.

Ships' flags and indicators

Ships carry a vast array of various types of flags with different colours and signs. These are used to indicate the ship's condition. Signal flags, or semaphore flags as they are sometimes known, have been used since the earliest days of seafaring and continue to be used to this day. In addition to flags, ships also carry and use various physical indicators to signal a ship's condition. Two of the most commonly used physical indicators are the black ball and the black diamond. The black ball is a daytime signalling shape used to indicate important information regarding the ship's condition. For example, a vessel at anchor will display a black ball at the fore-most end of the fo'c'sle. A ship not under command (NUC) will have two black balls in a vertical line hoisted on the highest mast. Alternatively, a black diamond is also a daylight signal which indicates the ship is unable to manoeuvre by itself or is under tow.

Celestial navigation

Celestial navigation, or navigation by the stars, is one of the oldest ways of navigating the vast oceanic expanse. To calculate and tabulate positions by way of celestial objects, it is necessary to understand the correlation between the angle of an object and its horizon with the vessel. To do this, we use a device called the sextant and reference manuals such as the nautical almanac. The longitudinal and latitudinal position of the ship can be determined within a relative scale of accuracy by comparing celestial bodies such as the sun and the moon, planetary objects such as Jupiter and stationary objects like the North Star. The ship's position is then corroborated after eliminating calculative errors. While the origins of celestial navigation were rudimentary, technological developments in the 1800s made navigation at sea much more accurate. Indeed, the use of celestial navigation was much in vogue until the latter part of the twentieth century, when computerised systems started replacing the ancient skills of yesteryear. Today, celestial navigation is predominantly the preserve of hobbyists and traditionalists, though it is still taught to maritime professionals as a form of redundancy.

Sextant

The sextant is an instrument that measures the angles between two or more objects at sea. The sextant derives its name from the Latin for 1/6th (Sextans) as the sextant is shaped to form a sector, which is 60 degrees of 1/6 of a circle. The sector-shaped part is called the frame. A horizontal mirror is attached to the frame, along with the index mirror, shade glasses (sunshades), telescope, graduated scale, and a micrometre drum gauge. The normal graduations of the arc, to the left of zero, extending from 0 to 130 degrees are referred to as ON the arc. To the right of 0 degrees, the graduations extend for a few degrees and are referred to as OFF the arc. When reading OFF the arc, graduations of the micrometre should be read in the reverse direction (59 as 1', 55 as 4' and so on). The sextant works by applying the principle of double reflection. As a ray of light (incident ray) enters the frame, it is reflected off a plane mirror. The angle of the incident ray is equal to the angle of the reflected ray when the incident ray and reflected ray lie on the same plane. When a ray of light suffers two successive reflections in the same plane by two plane mirrors, the angle between the incident ray and the reflected ray becomes twice the angle between the mirrors. This means the sextant can measure angles of up to 120 degrees, though, in actual fact, because the arc of the sextant is a little over 60 degrees, the total angle measurable by the sextant is closer to 130 degrees. The sextant is a vital piece of navigation equipment and can be used by the ship's navigator to determine the Vertical Sextant Angle (VSA), the Horizontal Sextant

Angle (HSA), and altitudes. Using the sextant requires a good deal of training and experience, but once mastered, it is an excellent tool that can be used when electronic navigation equipment and systems fail.

Figure 3.2 Quartermaster M.A. James using the sextant onboard *USS Vella Gulf.*

Navigational lighting

As with all modes of transportation, ships use lights during the hours of darkness. The use of lights at sea is so important it is often the first thing deck cadets learn when they begin their studies to become a navigation officer. The primary function of ships' lights – or navigation lights as they are correctly called – is to prevent accidents and collisions at sea. All watercraft, from small pleasure boats to giant oil tankers, are required by law to have navigation lights as part of their navigational systems. The system of navigation lights was introduced in 1849 by the United Kingdom following an Act of Parliament. In 1889 the International Maritime Conference was established by the United States to set formalised guidelines for the prevention of accidents and collisions at sea. In 1897, these rules were adopted internationally. The colour of the lights was those set out in the 1849 rules; that is red for port, green for starboard and white. These colours remain the same to this day. The way the navigation lights are set up on ships and boats is in accordance with the IALA Buoyage System (see Chapter 1). Part C of the COLREGS sets out the requirements for boats and ships to comply with the standards and rules pertaining to navigation lights but, in essence, the pattern can be described as: (1) there is a light at the right-hand side of the boat (right side when facing the bow of the vessel, which is known as the starboard side) and is green in colour; (2) there is a light at the left-hand side of the boat (left side when facing the bow of the vessel, which is known as the port side) and is red in colour; (3) both sidelights show an unbroken light over an arc of the horizon of 112.5 degrees such that from right ahead it can be viewed to 22.5 degrees on either side; a white light is also placed at the back of the boat (known as the stern side). This shows an unbroken light over an arc of horizon of 135 degrees and is fixed to show the light 67.5 degrees from right aft on either side; (4) the mast of the boat must also have night lights. The colour of this light is white. Two masthead lights must be in place, with the second one higher than the first, when the length of the vessel is greater than 50 metres (164 ft). This shows an unbroken light over an arc of the horizon of 225 degrees and is fixed to show the light from right ahead to 22.5 degrees abaft on either side of the ship. The visibility range of navigational lights varies between three to six miles. The lights that are used for ships and larger boats have a longer range of visibility when compared to smaller watercraft. Table 3.1 shows the visibility of lights as per COLREGS rule 22. All ranges are in nautical miles and the lengths are in metres.

Table 3.1 Visibility of lights as per COLREGS rule 22

	Range	*Range*	*Range*
	Length >50 m	Length >12 m <50 m	Length <12 m
Masthead Light	6	5 or 3	2
Side Light	3	2	1
Stern Light	3	2	2
Towing Light	3	2	2
All Round Light	3	2	2

The importance of navigational lights for ships cannot be stressed enough, but to understand their practical application, we should imagine a head-on situation (i.e., two vessels on a reciprocal course). In accordance with COLREGS rule 14, upon seeing each other's lights (called the 'aspect') both vessels are required to alter their

Figure 3.3 Navigation lights, Rhine-Herne Canal, Germany.

course to starboard so as to pass on each other's port sides. This means each vessel can pass the other without the risk of colliding.

Wind indicators

Wind indicators are used on ships and boats to determine the prevailing direction of the wind. It is also known as a Windex. Wind indicators use the same technology as traditional weathervanes, in that they rotate and point in the direction the wind is blowing. The main reason why wind indicators are used on ships is the unpredictability of oceanic winds and weather. Winds are easily influenced by large bodies of water. This, in turn, affects how ships behave when underway. Strong winds can blow ships off-course; hinder their progress resulting in burning more fuel; and can be an indicator of sea squalls and storms. Wind indicators also help sailors on smaller watercraft and yachts to know when to raise the sails on their boat. There are three main types of wind indicators that are in use today. These are: (1) Type I. This type of wind indicator is the most basic type and is known as the Luff Tell Tales (luff is the part of a ship's sail that is very close to the wind). Luff Tell Tales are generally attached to the sails rather than positioned separately to the ship's mast. This is problematic as the Type I, being attached to the sail itself, it is often obscured from vision; (2) Type II. These types of wind indicators are known as Shroud and Stay Tell Tales. They are better designed than the Type I wind indicator are more clearly visible by the ship's navigator: (3) Type III. This variety of wind indicator is by far the best out of the three types. They are the most modern and function electronically. This helps to reduce the margin of errors that can happen if the wind direction is monitored manually. The wind indicator is placed at the top of the mast. Unlike the Types I and II, the Type III wind indicator can also monitor the ship's speed of the ship together with the direction of the wind.

In this chapter we have very briefly looked at some of the main manual aids to navigation carried and used on board merchant ships. In the next chapter, we will turn our attention to virtual aids or electronics aids to navigation.

4   Virtual aids to navigation

Everybody, both onshore and at sea, has a responsibility for improving the safety of ships and for saving lives at sea. Aids to Navigation, also known as AtoN, are important tools that enhance ship safety. Unlike the roads and motorways that we drive on, the waterways do not have road signs that tell us our location or the distance to a destination, or of the hazards we may expect to encounter along the way. Instead, navigators are provided with aids to navigation. For hundreds of years, maritime authorities have marked safe and hazardous waters with buoys and beacons. In summary, aids to navigation help ship's navigators to find and safely navigate within a narrow channel in a very wide expanse of water.

A navigational aid, AtoN, or navaid, is any form of marker that guides the navigator towards and through safe waters and helps mariners to determine their position with respect to land or any navigational hazard or hidden danger. Traditionally, aids to navigation have been physical aids such as lighthouses, buoys, and beacons. The introduction of virtual aids to navigation has been one of the greatest achievements in maritime navigational history. Though many are presently under development, they are used by maritime administrations around the world. Importantly, a virtual aid to navigation itself does not physically exist – unlike buoys and beacons – but comprises a signal broadcast to a location in a waterway. It can be described as digital information transmitted from an Automatic Identification System (AIS) station located elsewhere for a specified location without itself being present. The basic symbol of a virtual AIS aid to navigation looks like a diamond shape with a cross-hair in the centre on Electronic Chart and Display Information Systems (ECDIS) or on radar. Virtual aids to navigation offer great potential in enhancing safety and their use brings navigators various advantages, some of which we will explore below. First and foremost a virtual aid to navigation can be used in locations and situations where it is not practical or possible to place a physical buoy, beacon, or lighthouse. In these situations, virtual aids to navigation can be used wherein an AIS coastal station can transmit information to mark the location of a hazard or safe water and therefore assist navigators in real time. Second, virtual aids to navigation can be deployed rapidly to mark immediate wrecks. The implementation of virtual aids to navigation can be completed within a very short time, thereby assisting navigators to avoid unexpected hazards and dangers to navigation that can arise at any time without any prior information by giving early warning. Third, virtual aids to navigation are not affected by the weather, climate, or time of day.

DOI: 10.1201/9781003291534-4

Unlike traditional buoys or beacons that may not be detectable in bad weather, virtual aids will always be obtained so long as the ship is able to receive ECDIS or radar overlay inputs. Fourth, virtual aids to navigation are particularly useful where physical buoys are seasonally lifted or misplaced due to swell or ice or when a buoy is off-station or damaged due to natural conditions. It is often difficult and prohibitively expensive to install and maintain physical buoys in areas prone to harsh environmental conditions. Virtual aids to navigation bypass this difficulty altogether. Fifth, unlike physical buoys and beacons, virtual aids to navigation are easy to install and do not require costly physical infrastructure. They also require less maintenance. This has made virtual aids to navigation increasingly popular with coastal maritime administrations. Sixth and last, virtual aids to navigation can be used to mark anchorages, restricted or dangerous areas as well as environmentally sensitive and isolated coastlines where there are no buoys or beacons. They can also provide additional information regarding reporting points to officers. Furthermore, they provide navigators with information well in advance which means the navigator can plan and adjust their course accordingly to avoid any potential collision hazards or dangers to shipping.

Although virtual aids to navigation have come to play a major role in enhancing navigational safety and bring many benefits to the modern seafarer, we must be vigilant when using them and recognise their limitations. We must consider the fact that positional data contained within transmissions may be inaccurate. AIS data are susceptible to spoofing or jamming. Furthermore, if the AIS unit is malfunctioning onboard the vessel, there is the risk the navigator may receive false data and thus might not be aware of the actual position of the vessel. GPS errors can cause positional inaccuracies. All in all, it is important for ship's navigators to remember virtual aids to navigation are just that – an aid to navigation. They must never replace the navigator's better judgement, nor should they ever take precedence over other established navigational techniques.

Automatic Identification System (AIS)

In the previous section we were introduced to the world of Automatic Identification Systems (AIS). AIS is often a confusing area to delve into, with many questions arising such as: what is AIS? Why do I need it? And what type of AIS does my ship need or have? In simple terms, an AIS is an automated tracking system that displays all other vessels in the vicinity of the ship. It consists of a broadcast transponder system that operates on the VHF mobile maritime band. The ship fitted with the AIS also appears on the screens of other ships within the same proximity. If AIS is not fitted or is not switched on, there is no exchange of information on ships in the area via AIS. Where AIS is installed, it must be switched on unless the master deems it necessary to turn the AIS off, for example, for security. The working mode of AIS is continuous and autonomous. Although the AIS is fitted on ships for the identification of ships and navigational marks, it is only an aid to navigation and should not be used as a means for collision avoidance. VTS also use AIS to identify, locate and monitor vessels within their area of supervision. AIS is also useful for other purposes. For example, the Panama Canal uses AIS to provide information regarding rainfall along the canal as well as wind in the locks. Legally speaking, the IMO has mandated through SOLAS regulation V/19.2.4 that all vessels of 300 gross tonnes and above, engaged on international voyages, and all passenger ships irrespective of size, must carry AIS onboard and keep it in an operational setting.

Types of AIS

There are two classes of AIS:

(1) *Class A:* mandated for all vessels of 300 gross tonnes and above engaged on international voyages as well as all passenger ships of any size; and
(2) *Class B:* which provides limited functionality and is intended for non-SOLAS-compliant vessels such as pleasure watercraft.

AIS operates principally on two dedicated radio frequencies or VHF channels: (1) AIS 1: which works on 161.975 MHz channel 87B (simplex, for ship-to-ship communications); and (2) AIS 2: 162.025 MHz channel 88B (duplex for ship-to-shore communications). AIS uses Self Organising Time Division Multiple Access (STDMA) technology to meet the high broadcast rate though the frequency is limited to line of sight, which is about 40 miles (64 km) or so.

How AIS works

Originally, AIS was used terrestrially, meaning the signal was sent from the ship to a shore-based receiver and had a range of approximately 20 miles (32 km) (taking into consideration the natural the curvature of the Earth). As ships began sailing further away from land, they began sending signals to low-orbit satellites; these then relayed that information to shore-based stations. This meant ships could sail far from land and still maintain contact with land authorities.

The AIS system consists of a single VHF transmitter, two VHF TDMA receivers, one VHF digital selective caller (DSC) receiver, and a standard marine electronic communications link to the shipboard display and sensor systems. Position and timing information is normally derived from an integral or external GPS receiver. Other information that is broadcast by the AIS is electronically obtained from shipboard equipment through standard marine data connections. Although only one channel is necessary, each station transmits and receives over two radio channels to avoid interference and communications loss from ships. A position report from one AIS station fits into one of 2,250 time slots which are established every 60 seconds. AIS stations continuously synchronise themselves to each other, to avoid an overlap of slot transmissions. AIS is relatively easy to install as AIS is integrated directly into the ship's bridge systems or multifunctional display.

Types of data transmitted by AIS

As we saw earlier, the AIS system transmits different types of data, including the following:

(1) Static Information (every 6 minutes and on request):

  (a) MMSI number
  (b) IMO number
  (c) Name and call sign
  (d) Length and beam
  (e) Type of ship
  (f) Location of position fixing antenna

(2) Dynamic Information (depending on speed and course alteration)

  (a) Ship's position with accuracy indication
  (b) Position timestamp (in UTC)
  (c) Course Over Ground (COG)

(3) Voyage Related Information (every 6 minutes, when data are amended, or on request)

  (a) Ship's draught
  (b) Type of cargo
  (c) Destination and ETA
  (d) Route progress (waypoints)

(4) Short safety-related messages

  (a) A free-format text message addressed to one or many destinations or to all stations in the area. This content could be such as buoy missing, iceberg sighting, etc.

AIS as a surveillance tool

In coastal waters, shore-based authorities may establish automated AIS stations to monitor the movement of vessels through their area of supervision. Coastal stations can also use AIS channels for shore-to-ship transmissions, to send information on tides, Notices To Mariners and local weather conditions. Coastal stations may use the AIS to monitor the movement of hazardous cargoes and control commercial fishing operations in their waters. AIS may also be used for Search and Rescue (SAR) operations enabling SAR authorities to use AIS information to assess the availability of other vessels in the vicinity of the incident.

AIS as an aid to avoid collisions

AIS contributes significantly to navigational safety. All the information that is transmitted and received via the AIS system enhances the effectiveness of navigation and greatly improves situational awareness and the onboard decision-making process. As an assistant to the Officer of the Watch, the tracking and monitoring of targets by the AIS as well as determining information on the closest point of approach (CPA) and time to closest point of approach (TCPA) makes the task of the ship's navigator just that little bit easier. However, it is essential officers do not rely on the information from the AIS for collision avoidance. AIS is only an additional source of information for the OOW and only supports the process of navigating the vessel.

Limitations of AIS

As with all navigational and/or electronic equipment, the AIS has its limitations. First and foremost, the accuracy of AIS information received is only as good as the accuracy of the AIS information transmitted; the position received on the AIS display might not be referenced to the WGS 84 datum; over-reliance on the AIS can cause OOW complacency and lead to mistakes; AIS users must be aware that erroneous information might be transmitted by the AIS from other ships; not all ships are fitted with AIS, therefore gaps in coverage and information might exist; the OOW must be aware that even if AIS is installed, it might be switched off without the OOW realising, thereby negating any information that might be transmitted from that ship. So, in summary, it is not prudent for the OOW to assume the information received via AIS is absolutely accurate; whilst it is prudent to use AIS as an aid to navigation, any information received should be treated with a healthy dose of scepticism and not absolutely relied upon.

Vessel Tracking System (VTS)

The Vessel Tracking System (VTS), as the name suggests, is an electronic system that allows authorities to identify, track and monitor a ship's position and course. There is a wide range of VTS available, though the most commonly used platform is GPS. GPS is a vessel tracking system works by transmitting signals from the target ship to high-orbit satellites. This means the system is highly effective and provides pinpoint accuracy. Although GPS is by far the most popular VTS, there are other VTS systems available including AIS, which we briefly discussed earlier. A vessel tracking system like AIS uses GPS technology to aid and serve as a transmission and receiving point using VHF radio channels. The AIS acts as a transceiver between the ship and the coastguard. There is also a vessel tracking system known as Ship Loc, which can be installed on board vessels and provides authentic readings pertaining to the location and position of the ship as well as ancillary information such as air pressure, wave pressure, wind direction and speed and a range of other useful data which are important for ship's navigation.

Vessel Monitoring System (VMS)

Vessel Monitoring Systems (or VMS) are typically installed on board commercial fishing vessels and allow compliance authorities to manage the activities of individual boats within the fishing fleet. VMS may be used to monitor vessels within territorial waters (the 12-nautical mile zone) or the Exclusive Economic Zone (up to 200 nautical miles). The purpose of the VMS is to improve the management and sustainability of aquaculture by preventing over-fishing, illegal fishing and protecting the marine environment.

Integrated Bridge System (IBS)

The Wärtsilä Encyclopaedia defines the Integrated Bridge System or IBS as "a series of interconnected and closely grouped screens and modules allowing centralised access to navigational, propulsion, control and monitoring information. The aim of IBS is to increase safe and efficient ship management by qualified personnel." In other words, it is a combination of systems that are interconnected to allow centralised monitoring of various navigational tools. IBS allows the acquisition and control of sensor information from several simultaneous operations, such as passage execution, communication, machinery control, and safety and security. In effect, IBS is a kind of navigation management system which links together all other systems on the ship's bridge to provide every detail pertaining to the ship's navigation in a single location. Because IBS is intrinsically linked to the type of ship and the type of operations the ship is designed to perform, every ship will have a different IBS system, albeit the basic principles are not dissimilar. The actual design and operation of the system varies according to the design of the ship's bridge, the various types of equipment used by the ship, and the general layout of the bridge's equipment interface.

Irrespective of design or function, where installed, the IBS should support a minimum of two or more of the following roles:

(1) Execution of passage
(2) Communications
(3) Machinery control
(4) Cargo operations
(5) Safety and security.

It is worth noting that IBS is not mandatory on ships. Its installation and design criteria are laid out by classification societies. For instance, the NAV1 class from LR and the W1-OC class from DNVGL are examples of class arrangements for IBS. The factors which determine the functionality of IBS includes bridge design, the types of equipment fitted and their location and placement on the bridge. IBS can be divided into four main components:

(1) Technical System
(2) Human Operator
(3) MMI (Man Machine Interface); and
(4) Operational Guidelines.

Within the first component, Technical System, there are nine primary elements consisting of:

(1) Autopilot
(2) Dual Radar/ARPA
(3) Gyro compass
(4) Position Fixing Systems

(5) Dual ECDIS Setup (Master and Backup)
(6) Conning Display (available at the conning position to show information that summarises the important navigational sensors on passages and in port; it also provides the OOW with a centralised location for monitoring sensors and console settings)
(7) Power Distribution System
(8) Steering Gear; and
(9) GMDSS.

In terms of safety and failure redundancy, SOLAS Chapter 5, regulation 19, paragraph 6 stipulates the

> integrated bridge system shall be so arranged that failure of one sub-system is brought to the immediate attention of the officer in charge of the navigational watch by audible and visual alarms and does not cause failure to any other sub-system. In case of failure in one part of an integrated navigational system, it shall be possible to operate each other individual item of equipment or part of the system separately.

For this reason, the IBS is fitted with various alarm systems, which are outlined below. As with all AtoN, it cannot be stressed enough that IBS is an excellent system for navigation, but officers on watch should never completely rely on the equipment and should always apply proper visual navigational watchkeeping techniques as the first resort. Furthermore, as IBS is a support system, and not a replacement system, proper guidelines should be provided in the bridge manual as to when to use and when not to use the IBS.

IBS alarm system

An alarm system links all the above-mentioned systems and provides an audio and visual signal in case of an emergency condition. On most ships, an additional alarm connected to the IBS is also fitted in the day offices and cabins of the navigational officers. This alarm provides a signal in the cabin within 30 seconds in the event that the OOW fails to acknowledge the alarm on the bridge. The alarm system also includes a watch safety or fitness alarm to monitor the alertness of the OOW. Several alarm acknowledgements points, each with a pre-warning alarm, gives the OOW notice that the alarm is about to activate. These are positioned around the bridge. If the OOW fails to acknowledge a navigational alarm, or if the fitness time interval expires, the alarm automatically sounds alerting the other navigation officers.

IBS power supply

The guidelines laid out for IBS state that should the IBS be subject to an orderly shutdown it must restart and condition itself in the default operational state. If the shutdown is unplanned or inadvertent, the IBS should, upon restart, be restored to full functionality with the same configuration in use prior to the shutdown. To enable an uninterrupted power supply, the IBS is powered from the ship's main power supply as well as emergency power supply, with automatic changeover from main to reserve supply. The critical parts of the IBS are also connected directly to a reserve power source which automatically kicks in in the event of a ship-wide blackout.

IBS and INS

Integrated Navigation System (or INS) is a combination of navigational data and systems which are interconnected to enhance the safe navigation of the vessel. IBS interconnects various other systems along with the INS to increase overall management efficiency.

Bridge Navigational Watch and Alarm System (BNWAS)

The Bridge Navigational Watch and Alarm System, often abbreviated to BNWAS, is an automatic system which triggers an alarm if the OOW falls asleep, becomes otherwise incapacitated on watch, or is absent from the bridge for a long time. The BNWAS is automatically engaged when the ship's autopilot is activated. On passenger vessels, the IMO stipulates the minimum requirement for BNWAS is a dormant stage and three alarm stages though on cargo vessels, the second the stage alarm may be omitted. For Stage 1, when the autopilot is engaged, the OOW is required to signal their presence to the BNWAS every 3 to 12 minutes. The OOW does this by moving an arm in front of a motion sensor, pressing a confirmation button, or directly applying pressure to the BNWAS centre. The OOW is alerted to the alarm by a flashing light. If a confirmatory signal is not received by BNWAS within 15 seconds, a Stage 2 alarm will activate on the bridge. The OOW then has a further 15 seconds to acknowledge the alarm before a further Stage 2 alarm triggers in the master's and chief officer's cabins. The second Stage 2 alarm can only be cancelled by either the master or the chief officer's presence on the bridge. If neither the master nor the chief officer cancels the alarm within a specified time period (usually between 90 seconds and three minutes depending on the size of the vessel) a Stage 3 alarm will sound throughout the vessel. This then alerts other personnel that there is potentially an issue on the bridge that requires investigation. In addition to monitoring the OOW, the BNWAS may also be used in an emergency situation, whereby the bridge personnel can activate a Stage 2 or Stage 3 alarm to summon assistance. The IMO has established requirements for vessels to carry BNWAS under the SOLAS regulations, namely under the amendment of 5 June 2009, which came into force on the following dates for ships categorised by size: (1) July 2011: new vessels in excess of 150 tonnes; (2) July 2011: all passenger vessels regardless of tonnage; (3) July 2012: (4) all vessels in excess of 3,000 tonnes; July 2013: (5) all vessels between 500 and 3,000 tonnes; (6) July 2014: all vessels between 150 and 500 tonnes.

Autopilot

In days long past, the Quarter Master was an essential member of the bridge team. It was a title bestowed upon experienced and trusted able-bodied seaman whose job was to steer the ship in accordance with the master's and officer's helm orders. The Quarter Masters kept watches and took turns to man the helm. This practise continued into the first half of the twentieth century before automation took over and electronic navigation systems negated the need for human intervention. During the 1920s, automated steering and helm control systems were gradually introduced on board merchant ships. Today, that system is called the autopilot, and is considered one of the most advanced and technologically sophisticated navigation tools used on ships. The autopilot is synchronised with the gyro compass to steer manually inputted course data, which are then cross-referenced with the gyro heading. The autopilot steers the ship according to the course that is manually entered by the ship's navigator. The system controls the steering gear which turns the rudder.

Furthermore, modern autopilot systems can integrate and synchronise with ECDIS, which allows the ship's navigator to set a manual course once using the electronic chart, after which any course deviations are automatically picked up and actioned by the autopilot. Compared to the days of manual steering, the invention and advancement of the autopilot has revolutionised the maritime industry. Yet, like all things, over-reliance on the equipment and poor comprehension of its efficiencies and limitations can lead to avoidable accidents. Good training and experience of the autopilot system is needed to really understand and use the technology to its fullest potential. Listed below are some of the most important points navigators need to consider when using the autopilot in place of manual navigation and steering.

Figure 4.1 Modern ship's bridge, *Norwegian Jade*.

Rate of turn and rudder limits

The method of turn is the most important control of the autopilot system. The system will use the selected turn method to action course alterations. The user can input the limit of such turn methods, which are as follows:

(a) *Rate of Turn.* This is the most used turn method. With this method, the user can set a value of rate of turn between 1 and 300 degrees (this varies between different models). When turning, the rudder will move as much as it takes to attain the required turn rate without exceeding the set value. This means the navigator must consider the vessel's manoeuvring characteristics and set a value that is safe for the vessel according to the prevailing conditions.

(b) *Rudder Limits.* The rudder limit method allows the user to set a value from 1 degree to the maximum rudder angle. With this method, when altering course, the rudder will not exceed more than the set limit. Again, the vessel's manoeuvring characteristics must be considered when setting the rudder value.

(c) *Turn Radius.* Modern systems allow turning by radius as well. With this method, the user can input the turn radius in nautical miles.

Steering gear pumps

Steering gear pumps are used to pump hydraulic oil to actuate the steering gear unit (the ram) which in turn moves the rudder in the required direction. That means, when more pumps are running, the rudder will move more swiftly. The number of pumps available varies as per the steering gear unit. The ship's navigator should be aware of the pumps and use them accordingly. When operating the autopilot in areas with high traffic density – where sudden and swift alterations may be required – the maximum number of steering gear pumps should be running. In ocean cruising and open sea navigation, where there is likely to be much less traffic, the number of pumps running should be reduced to the appropriate minimum.

Off-course alarm

An off-course alarm serves the purpose of notifying the ship's navigator of any difference or discrepancy in the set course and the actual heading of the vessel. The navigator can manually set the required number of degrees after which the alarm will sound to notify the navigator that the set degree of difference has been exceeded. However, the navigator must maintain a ready check on the course changes as in some cases when the gyro compass wanders its course, the autopilot will follow the wandering compass and fail to sound the alarm.

Manual mode

The steering controls of the system can be used in an Automatic or Manual mode. This means the ship can navigated manually by the OOW or automatically by the system. In Manual Mode, the vessel can be hand-steered by using the Follow-Up Helm or a Non-Follow Up Emergency Tiller. Hand steering is used when the ship is manoeuvring, navigating in restricted waters, channels, and any other area areas with high traffic density. The Non-Follow Up tiller, when used, moves the rudder in a desired direction, but not to a specific angle. This is used in the event of emergencies. Subsequently, the navigator and or OOW must be fully familiar with the procedure of switching from automatic to manual modes and vice versa.

Traffic density

The use of the autopilot is not recommended when navigating in areas with high traffic density, in narrow channels and through traffic separation schemes and other restricted waters. The autopilot may not be efficient enough to turn the vessel spontaneously while navigating in such areas demanding swift alterations and manoeuvres to avoid a collision or close-quarters situation. If the autopilot is used in such cases, all steering gear pumps must be switched on for enhanced rudder response.

Speed

The autopilot system works inefficiently at reduced speeds. The use of the autopilot is not recommended when the ship is manoeuvring or steaming at low speed. The system allows the users to synchronise with the Speed Log to receive feeds on the ship's speed. The navigator should maintain a constant check on the Speed Log as any error in the log speed will manifest in the autopilot system. The system also allows the navigator to manually input the desired speed; when doing so, however, it is important to set a value as close as possible to the actual speed of the vessel.

Weather conditions

Rough weather and heavy sea conditions have adverse effects on the performance of the autopilot. Uncontrolled yawing of the ship can result in excessive rudder movement. Modern autopilot systems have a Weather Control option whereby the system automatically adjusts the setting to adapt to the changing weather and sea conditions. It also provides an option for the navigator to manually set a specific value.

Gyro compass

The autopilot system is functionally dependent on the gyro compass. If there is any error or fluctuation in the gyro heading, there will be an equivalent change in the course steered by the ship. In the worst-case scenario, when the gyro fails, the system will lose track of its heading and will be unable to steer the in required course. In an emergency, power blackout or gyro failure, the system should be immediately changed over to Manual mode. The helmsman must use the helm to steer the course using the magnetic compass.

Important alarms and signals

In addition to the off-course alarm, the autopilot must also be integrated with the following alarms and signals:

(a) Failure or Reduction in Power alarm: which will sound in the event of auto-pilot failure or in the event of a reduction in the power supply to the heading control or monitoring system

(b) Sensor Status Monitoring: if any of the sensors in the autopilot system fails to respond, it should indicate the failure with an audible alarm

(c) Heading Monitor: where the ship is required to carry two independent compasses, a heading monitor to track the current heading information by independent heading sources must be provided. An audio-visual alarm must be provided if the heading information in use diverts from the second heading source beyond a set limit. It should also be provided with a clear indication of the actual heading source.

Important limitations

The autopilot system must be such that the pre-set heading cannot be altered by the intentional intervention of onboard personnel and the heading control system should change the course to pre-set heading without overshooting its position. As we have stated above, the autopilot is undeniably a major improvement in maritime navigation. But it is the responsibility of the navigation officers to ensure they are fully trained and competent in its use, and recognise the limitations of the autopilot system, and, more importantly, recognise when and how to switch from automatic steering to manual steering. Although autopilot systems will vary from ship to ship, the basic working principle and features will be the same. Even so, navigation officers joining their ship for the first time are strongly advised to acquaint themselves with the autopilot manufacturer's instructions before attempting to operate the system for the first time.

5   Marine radar

The marine radar is the piece of equipment on the ship's bridge that is perhaps most used by the OOW when carrying out a safe navigational watch. A mandatory aid to navigation, the radar is used for the identifying, tracking (with integrated ARPA) and the positioning of vessels (including one's own vessel). The marine radar is classified either under the x-band (10 GHz) or S-band (3 GHz) frequencies. The x-band, being of higher frequency than the S-band, is used for a sharper image and better resolution whereas the S-band is predominantly when visibility is poor such as during heavy rain or fog or when vessel identification and tracking is required. Ship-tracking devices are compulsory under the COLREGS. SOLAS Chapter 5, regulation 19 states that

> all ships of 3,000 gross tonnes and upwards shall, in addition to meeting the requirements of paragraph 2.5, have a 3 GHz radar or where considered appropriate by the Administration, a second 9 GHz radar, or other means to determine and display the range and bearing of other surface craft, obstructions, buoys, shorelines and navigational marks to assist in navigation and in collision avoidance, which are functionally independent of those referred to in paragraph.

With the help of marine radar, accidents can be prevented at sea by determining the closest point of approach (CPA) and the time to closest point of approach (TCPA), electronic bearing line (EBL), variable range marker (VRM), and so forth. The benefit of radar does not stop when the ship is no longer at sea. Even alongside or within port confines, radar can be used to monitor traffic at short range. The marine radar has a screen (referred to as the Plan Position Indicator (PPI)) which displays all the targets that are present within the radar range. Since these objects are clearly visible on the screen, navigating and monitoring the position of the ship is vastly improved.

The term "radar" is an acronym for Radio Detection and Ranging. The marine radar works on the basic principle of electromagnetic waves. The radar antenna sends the high-speed electromagnetic waves to establish the location of a target, which is the distance, the velocity, and the direction the wave travelled together with the altitude of the object, whether moving or stationary. Electromagnetic energy travels through the air at a constant high speed, equivalent to the speed of

DOI: 10.1201/9781003291534-5

Figure 5.1 Typical ship's radar arrangement, *Volcan de Taburiente*, Tenerife, Spain.

light, or 186,411 miles (300,000 km) per second. The object may vary from ships, boats, terrain, weather formations, coastal formations, buildings and so forth. The radar system sends out electromagnetic waves as a high-speed signal which travels several miles in the same direction the radar is facing. If there are no objects in the direction of the wave, the radar screen will be blank. If there is an object, this will reflect the wave and bounce it back to the radar receiver. Once the ship has received the returned signal, the onboard radar computer calculates the distance between the ship and the object along with its location. Subsequently, radar provides three critical pieces of information: the location of an object, the range of the object, and the direction the object is travelling.

To determine the position of the object, the radar antenna continuously rotates 360 degrees at the highest point above the bridge. Typical antenna rotation rates are anywhere between 24 and 45 revolutions per minute, meaning a complete rotation can be achieved every 1.3 to 2.5 seconds depending on the system used. Because the antenna is positioned high above the ship's superstructure, and rotates completely, it can paint an accurate image around the entire circumference of the vessel.

How marine radar is used

Modern marine radar has many uses and has certainly come into its own as a necessary aid to navigation.

Amongst its many uses, marine radar can be utilised to: calculate the range and bearing of a target and thereafter use that information to determine an appropriate speed and course; integrate with other shipboard equipment (such as ECDIS) to provide accurate and up to date information; navigate own vessel in accordance with prevailing traffic and make necessary collision avoidance manoeuvres; fix the ship's position using terrestrial objects such as lighthouses, maritime markers and buoys; differentiate between multiple targets in high traffic density areas; determine weather conditions (to a limited extent); liaise and communicate with VTS in controlling coastal traffic; assist in the use of parallel indexing; alleviate some of the manual workload incumbent upon the OOW and other bridge watchkeepers; and, finally, to greatly assist the pilot during pilotage operations. Lastly, one major advantage of marine radars over earlier models is that the power and electricity consumption used by them is vastly improved. This means that the marine radars are not just user-friendly but also reduce the amount of power needed for marine radar systems to function. This provides additional saving benefits to shipowners and operators, and the wider environment.

How radar and ARPA complement each other

Marine radars have played a critical role in ship navigation for over six decades and will continue to do so for as long as humans set sail. Radar technology has improved immensely from its humble beginnings and the application of computer technology to commercial marine radar sets has resulted in the introduction of ARPA. ARPA provides all the necessary information for the navigator through a simple to read interface. Collision avoidance and detection data is thus readily available to the radar user in real time, and literally at the click of a button. Before we go into detail about radar watchkeeping, it would be useful to first look at how the radar system works. We have already covered some of the basic principles of electromagnetic waves, and how they bounce or reflect of objects, but what of the actual components that make up the radar system? Radar, or radio detection and ranging, consists of various parts which can be broadly divided into one of four categories: (1) the transmitter; (2) the receiver; (3) the scanner; and (4) the display.

The working principle of radar starts with an oscillator known as a magnetron, which has a fixed frequency. The magnetron takes electrical inputs from a power source, passes it through a delay line and then through a modulator. The modulator produces an output of electromagnetic energy called a pulse. These pulses are sent to the scanner or the antennae through a metallic tube known as the wave guide. From the scanner, the pulse is sent out into the atmosphere. The number of pulses sent out through the scanner per second is called the Pulse Repetition Frequency. The pulses travel through the atmosphere at the speed of light and are reflected after striking any target in their way. The reflected echoes return to the scanner, after which the receiver processes and amplifies the echo. This then appears on the display screen as a 'blip', which is identified as the target. In addition to the components listed above, radar systems also use what is called a TR Cell which blocks the receiver part of the waveguide during the transmission and again blocks the transmitter end of the waveguide when the pulse is received. By doing so, the TR Cell avoids detecting the ship which is transmitting and receiving the signal. In addition to the TR Cell, the radar system also contains a mixer. The mixer works by filtering out echoes at the same frequency as local oscillations. Again, this prevents the radar from picking up the ship's own signal.

Working of ARPA

The ARPA is a computerised additional feature that complements the standard radar system. ARPA works by taking a feed of the ship's course and speed, and the target's course and speed, and then calculates the appropriate collision avoidance data. This means the ship's navigator is freed from trying to calculate these data themselves.

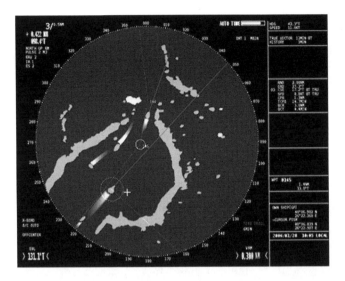

Figure 5.2 Typical radar set with multiple blips.

SOLAS requirements for radar and ARPA

SOLAS Chapter 5 sets out the carriage requirements for radar and ARPA on board ships. In the simplest terms:

(1) All ships of 300 gross tonnes or more and all passenger-carrying vessels must be fitted with 9 GHz radar and an electronic plotting aid
(2) All ships of 500 gross tonnes and above must be fitted with an automatic tracking aid to plot the range and bearing of other targets
(3) All ships of 3,000 gross tonnes and above must carry a 3 GHz radar or a second 9 GHz radar which are functionally independent of the first 9 GHz radar as well as a second automatic tracking aid to plot the range and bearing of other targets, which is functionally independent of the first electronic plotting aid.

It should be noted that although SOLAS does set out provisions that allows the use of any other equipment which can perform the same functions of radar and ARPA, in practicality there is no other equipment eminently suitable for this purpose. In essence, radar and ARPA are the only legitimate systems permitted onboard that fully comply with the SOLAS and COLREGS requirements.

Figure 5.3 Radar screen onboard the RV *Thomas G. Thompson*, Seattle, Washington, USA.

Principles of radar watchkeeping

Now that we have acquainted ourselves with some of the basic principles of what radar is and how it works, we can turn our attention to the practical business of radar watchkeeping. Radar watchkeeping is the process of monitoring the radar and making use of all its functions to make a full appraisal of any situation and early detection to avoid the collision of ships. Monitoring radar is not limited to one single observation, but multiple observations, of range and bearing. This provides a fuller and more accurate picture of the targets course and speed, which can then be analysed in relation to the course and speed of one's own ship. After a series of quick observations, the target should be plotted, and the target's data checked. The longer the plotting period, the more accurate the plot will be. Best practice dictates that a target should be plotted and monitored until it has fully passed and is clear from your own vessel. As mentioned earlier, the most effective way to use radar is to detect the target as early as possible (using range scales) and to plot it continuously as it approaches and passes your own vessel. When a target is detected well in advance, the actions to be taken by your own ship to avoid a potential collision course will be a lot easier. It will also help to avoid close-quarter situations and oversized altera- tions to avoid last-minute collisions. When it is necessary to have multiple targets in any one situation, for example, when passing through fishing grounds, it is advisable to tackle each situation as they arise rather than trying to manage every scenario at once. This means evaluating the risk of collision from each target and assigning a risk rating. Those targets which pose the most immediate risk must be tackled first. A fast-moving yacht that is 12 nautical miles off the starboard bow potentially provides a lesser hazard than a slow lumbering tanker which is 15 nautical miles off the port stern. Because the yacht is moving faster, and has greater agility com- pared to the tanker, the tanker poses the greater threat of collision. Targets with a higher risk of collision should be avoided before assigning importance to others. In summary, to avoid a collision incident, the radar and ARPA should be used as a navigational aid, and any actions and course alterations should be made according to good seamanship and in compliance with the COLREGS.

Basic radar settings and configurations

All deck officers, navigators and the OOW must have sound knowledge of radar and ARPA. The full advantage of having such tools and systems on board can only be if the personnel using them have the right knowledge and experience. Therefore, the settings and configurations of radar and ARPA should be thoroughly familiarised at the earliest possible opportunity when joining a ship for the first time. The OOW should be able to set up and configure the radar settings appropriate to their needs. As such, some of the basic – but important - radar controls are:

(1) Clutter Controls, which manages rain, gain and sea
(2) Pulse Controls and Range Controls
(3) Performance Monitor; and
(4) Manual Tuning.

With radar watchkeeping, it is essential to understand the limitations of the radar and ARPA system as well as their benefits. Over-reliability on radar and ARPA has been a cause of many accidents at sea. Radar users should be mindful of the fact that radar and ARPA are electronic systems incapable of independent rational thought. This means it is equipment that has its own limitations, the accuracy of which is largely dependent on the performance standard of the equipment. Subsequently, it is vital navigators and the OOW check the performance of the radar on an ongoing and regular basis. Some of the most important limitations of radar include:

(1) Small vessels, ice packs, and other small floating objects which may not be detected by the radar system due to their size or profile in the water
(2) Targets in the blind sector and shadow sector of the radar will likely not be accurate if displayed at all
(3) Range discrimination – this occurs when two small targets are on the same bearing and there is only a slight difference in range resulting in only one target being identified
(4) Barring discrimination – this is caused by two small targets on the same range but with a slight difference in bearing resulting in only one target being identified; and
(5) False echoes. Despite these limitations, the benefits of radar cannot be underestimated or oversold.

The OOW duties when handling the ship's radar

The OOW should be familiar with the workings of all navigational equipment used on board their ship. They should also be aware of the procedures for diagnosing and resolving minor faults in the event the system fails. First and foremost is the marine radar, which is without equal one of the most important navigational tools used on the ship. It should be kept running all the time and must undergo periodic tests to check for any operational errors. However, there are certain points that the OOW should bear in mind when operating the marine radar, including:

(1) The OOW should keep a close eye on the ship's course as occasionally small vessels, floating objects, and ice may not be readily detected due to their low profile
(2) Shadows and blind sectors should be considered and factored in when operating the radar;
(3) The plotting of targets should be done at long range wherever possible
(4) Use multiple plots to increase accuracy
(5) The detection of small targets is better achieved at shorter ranges
(6) The OOW should know how to handle video-processing techniques
(7) Long-range fixes should be used to acquire advance notice of approaching vessels, ports, and navigational hazards
(8) When deciding the range scale, two factors should be kept in mind: marine traffic in the area and the ship's speed
(9) Operational radar training should be practiced regularly in clear weather as this helps the OOW to build a better understanding of radar observations and target vectors during restricted visibilities
(10) The OOW should consider replotting or rechecking each target when there is a change in the ship's speed or course
(11) The OOW should maintain a regular watch on the variable range marker, fixed range lines, and electronic bearing lines
(12) The OOW must know how to use clutter control to avoid clogging the radar interface with unnecessary clutter
(13) The heading marker should be properly aligned with the fore and aft of the ship and with the compass heading
(14) The parallel index lines should be set correctly
(15) The OOW should regularly check for any gyro errors and confirm the accuracy of the heading line arrangement
(16) If a performance monitor is fitted, the quality of performance should be checked at regular intervals
(17) The OOW must check the identity of any fixed objects as they are identified; and lastly
(18) When deciding the CPA to avoid a collision, factors such as course, speed, and the aspect of the target should be considered.

Whilst this list is not exhaustive, it underlines the complexity of radar watchkeeping and hopefully exemplifies the importance of ongoing training to maintain radar watchkeeping competency.

Effective radar techniques

In the previous section, we discussed some of the main duties of the OOW when performing radar watchkeeping. In isolation, the points above probably seem abstract without any further context. To provide some context, we will now discuss some of the effective techniques the OOW can apply during their radar watch-keeping. Radar has been around for over half a century, and whilst today it is considered a vital piece of navigational equipment, this was not always the case. In the earliest days, experienced officers would often laugh and sneer at their junior colleagues for using radar, often preferring the old and tested method of manual plotting sheets and chinagraph pencils. A process that inevitably tested the patience of even the most seasoned sailors! Manual plotting also made tracking multiple targets almost impossible. Fortunately, today all ships are fitted with some variety of radar, and most are supported by ARPA. This means when the radar is not being used for coastal navigation, it can be used for collision avoidance. The regulations pertaining to the proper use of radar for collision avoidance are contained in the COLREGS, rules 6, 7, 8 and 19; and preceded by rule 5 which states that all available means must be used for keeping a proper lookout, which includes the use of radar. Furthermore, the requirement for ships, under rule 9, to keep as near as possible to the starboard side of a narrow channel or fairway also necessitates the use of radar. But, having radar on board and using it to best effect are very different sides of the same coin. To that end, we will briefly examine some of the areas where radar can be employed most effectively.

Conditions at sea are ever-changing in part because of weather conditions and in part due to variations in vessel traffic density; therefore, the ship's navigator must adjust the radar setting to their specific requirements and regularly monitor the settings throughout the watch. This is commonly referred to as tuning the radar. For example, if it was raining during chief officer's watch and the sky is clear during your watch, you do not need to apply the rain clutter. This is applicable for other radar controls too. It is often overlooked that improper use of clutter controls can mask targets or hamper their tracking. Similarly, the range scale depends on traffic conditions and the types of traffic being encountered as well as your own ship's speed. Some radars have pre-set settings depending on the type of the situation and anticipated use. Always ensure you are aware of such settings and use them appropriately. If pre-set settings are assigned names, that is perfectly well and good. A problem may arise, however, if pre-sets are assigned number – for example, Image 1 – as Image 1 on your previous ship's radar may be completely different to Image 1 on your present radar. Ship owners generally like to have uniformity of equipment on all ships across their fleet. This minimises maintenance costs and means personnel can transfer from one vessel to the next without needing retraining in specific systems. Like all marine navigation electronics, radar can fail, and you will need to know how to reset the equipment when needed. Most ships nowadays do not carry radio officers, and many have dispensed with the electro-technical officer. This means deck officers must know where the radar spare parts are stored and be ready to carry out minor maintenance tasks such as changing the magnetron. Another setting to be cautious of is inputting the vessel's speed through water (STW). This setting would be ideal if the only concern was collision avoidance. By tracking a vessel's speed, we can calculate the point of intersection and manoeuvre accordingly. The problem arises from the fact that radar is also used for situational awareness during coastal navigation and for ascertaining the set and drift speed over ground.

By monitoring the STW only, by applying the pre-set, this other critical information is easy to miss.

Most bridges are fitted with at least two marine radars. Generally, one of the radars will be an X-band radar and the other will be an S-band radar. Ensure you know the difference between the two and what settings are suitable according to each situation. It is never wise to navigate with two radars using the same range scale on both. It is very easy to fall into the bad habit of conning the vessel from one radar location. This can lead to an overdependence on one radar, even if the other radar is more suitable to the situation at hand. Due to their operational frequencies, X-band radars provide a clearer viewing screen and are best used for target detection and collision avoidance. S-band radars, on the other hand, are more suited to navigating through rain and fog as S-band radio waves have better penetration qualities. An important point to bear in mind, however, is that Search and Rescue Transponders (SART) are detectable only by X-band radar as they operate on a 9 GHz frequency. Remember this fact during watchkeeping as you may be called upon to save someone's life at sea. Insofar as frequencies are concerned, the same principles apply to radar beacons or RACONS and similarly with radar markers (RAMARKS). The SMS of some companies require that SART is switched on and thrown overboard when there is a person overboard (MOB) situation. In this instance, it is critical the X-band radar is monitored continuously for even the smallest location indicator.

One of the major limitations of marine radar is that they are often unable to detect objects due to their size, proximity to the sea surface, or when the object is a poor radar target because of its design or construction. Many naval vessels, for example, are designed specifically not to reflect radar signals. This also happens with wooden fishing boats as they have a poor reflective surface. Small unlit boats engaged in fishing are particularly difficult to detect at night, although spotting them during daylight hours can be equally as arduous due to the height and viewing angle from the bridge or the reflection of the water's surface. Therefore, when navigating in areas known to have such craft operating, it is good practice to assign crew members as additional lookouts. The best location is on the fo'c'sle deck. Every navigator must be conscious of the blind sectors and shadow sectors of their radar equipment. Notices advising the exact locations of these sectors are mandatory and should be posted near the radar on the bridge. Moreover, the navigator should also bear in mind that the radar may pick up many other forms of interference. These can include false echoes, second trace echoes, sea returns, clouds, low-flying aircraft and even large shoals of fish. Sometimes the design of the ship can provide unwanted interference or prevent a clean radar sweep. Ships with deck-mounted cranes or derricks, for example, often have their radar scanners installed on the forward mast. Likewise, cruise ships and even ultra-large container ships may have their radar scanners positioned on the after mast as their forward bridge design and large size makes it difficult to see clearly astern. Where a radar scanner that is positioned forward or aft is used for manual position fixing, remember this will give the position of the bow or stern of the vessel and not the navigation bridge; keep this in mind when navigating through waters with restricted sea-room.

As a final word of caution, bear in mind that natural occurrences such as floating seaweed can give very good radar returns due to their highly reflective surface.

Because radar only provides a two-dimensional picture, there are occasions when targets can be obscured due to an intervening echo. For instance, a target passing underneath a bridge will not be visible on radar as it will be masked by the echo of the bridge structure. Likewise, small targets can be overshadowed by islands, other ships and even buoys. Overhead cables can also result in false echoes. If there is any doubt whatsoever as to whether the target exists or not, it must be investigated. Out at sea this is not so much of an issue as there is theoretically plenty of space to manoeuvre, but in restricted waters, it is very likely you will not have the luxury of being able to outmanoeuvre a target that suddenly appears out of nowhere.

On some occasions, the AIS signal may also be adversely affected rendering it unreliable. Overhead cables, for example, are known to result in AIS targets showing odd courses and incredible speeds. Whenever using parallel indexing techniques, always remember that even though it is a means of near-continuous monitoring of vessel progress it still needs to be backed up by regular position fixing. It must be borne in mind that radar ranges are more accurate than radar bearings due to beam width and radar ranges and bearings must preferably be taken as far from the centre of PPI as possible by adjusting the radar to an appropriate range scale. Acquiring a range and bearing obtained from a single point will never be accurate enough; therefore, it is always preferable to get three lines of position (LOP). The best option will be to acquire three radar ranges with a good angle of cut, though this may not always be practical and two ranges along with one bearing will provide an adequate fix.

Adjusting the radar settings to personal preference

The radar is one of the most used equipment systems on board ships. It is designed for detecting and tracking targets at a considerable distance. Subsequently, it is of great practical value to every ship's navigator. The proper use of radar and radar plotting aids in both restricted visibility and clear weather can help prevent collisions and safeguard the integrity of the ship. Accidents unfortunately can and do occur when the OOW is not fully conversant with the operation of the navigation equipment onboard. For best use and reliable interpretation, it is essential that the radar operating controls are adjusted according to each user's preferences. In this section of the chapter, we will discuss how to interpret and understand the radar display screen.

Range measurement

Appropriate range scales should be used depending on the prevailing circumstances and the conditions of the environment the ship is in. Where two radars are used, one radar should be kept on a longer-range scale to obtain advance warning of the approach of other vessels, changes in traffic density, and proximity to the coastline. The other radar should be used as a short-range scale, to help detect smaller targets

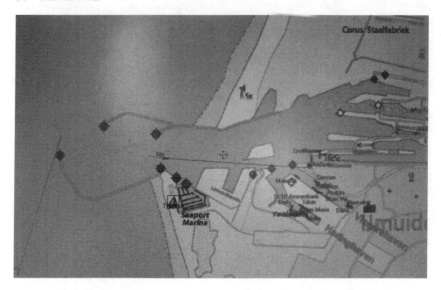

Figure 5.4 Radar screen with AIS connection to the Port of Ijmuiden, Netherlands.

more easily. This can be achieved by using the RANGE key on the keyboard to select the required range. The (+) key increases the range and the (−) key decreases the range.

Bearing measurement

Electronic Bearing Lines (EBL) are used to take the bearing of a given target. The EBL extends from your own ship's position to the circumference of the radar screen. If the bearing remains constant with a decreasing range, a risk of collision exists.

Gain

If rain or sea clutter is set too low, targets will be hidden in the clutter whereas if set too high, this can cause targets to disappear from the radar screen altogether. The radar may also detect rain, snow, or hail clutter in the same way as normal targets. The A/C RAIN and A/C SEA control is used to adjust the rain and sea clutter respectively. The scroll wheel is rolled clockwise or anticlockwise to increase or decrease the clutter.

Off-centre display

With off-centre display your own ship's position can be displaced to expand the view field without switching to a large range scale. However, when doing so, care should be taken to ensure at least one mile of viewing range is kept on the aft of the ship to view targets on the ship's aft or ships trying to overtake your own vessel.

The cursor is put to the position where you wish to move the ship's position to; then press the OFF-CENTRE key on the keyboard.

Target trails

Target trails can be particularly useful to the ship's navigator when making an early assessment of a situation. The trail can be either relative or true. A relative trail shows the relative movement between your own ship and the target. A true trail, however, shows the true target movements depending on the targets over the ground speed and course. Relative trails give an early indication if a close-quarter situation is developing, or if there is a potential risk of collision. Relative trails, when combined with true vectors, provides a fairly accurate indication of the relative movement of other vessels and the risk they present. The trail time can be adjusted according to specific or individual requirements.

Parallel index lines

Parallel indexing (PI) is a useful method of monitoring cross-track tendency. It helps in assessing the distance at which a given ship will pass a fixed object on a particular course. The index line is drawn parallel to the planned ground track and should touch the edge of a radar echo of a fixed object, at a range equal to the desired passing distance. Any cross-track tendency (such as those caused by a tidal stream, drift or current) will become apparent as the target moves off the parallel line. This technique can be used in both relative and true motion. To set the PI, use the trackball to select the PI line number box. Select a PI line number and push the left button to turn it on or off. Then, roll the scroll wheel to adjust the PI line orientation between 000 T to 359.9 T.

Heading, speed and course

The top right corner of the radar screen display shows the heading, speed, course, and speed over the ground, own ship's position, and the source. Speed can be entered from a log (STW), GPS speed over ground (SOG), or manually. The SOG is the speed of the vessel referenced to the surface of the earth. STW is the speed of the vessel referenced to the water in which it is navigating. In general, STW is used for radar collision avoidance to provide a more accurate indication of the target's aspect and SOG is used for navigational purposes. To activate the heading, speed, and course function, right-click the speed box to select the source for speed. Right-click the own ship position box to select the source of position data. This will be either GPS 1/2 or DEAD RECKONING.

Brilliance

The brilliance of the screen can be adjusted according to lighting conditions using the BRILL KEY. Turn the key clockwise or anti-clockwise. The brilliance box at the bottom left corner of the screen provides various palettes and other options.

Select the item needed and roll the scroll wheel to adjust the brilliance. The brilliance menu can be seen by right-clicking on the brilliance box.

Watch alarm

The function of the watch alarm is like DP TWAS. The watch alarm sounds the audio alarm at selected time intervals to help keep a regular watch of the radar picture. The countdown starts from the value set. Navigators often need to spend time inside the chart room, sometimes forgetting to keep a radar watch. The watch alarm can, therefore, be used to remind the navigator to return to the radar. On the radar screen, the ALARM 1 and 2 can be used to set the alarm. The ALARM ACK key should be pressed to silence the alarm.

Vector mode

Target vectors can be set Relative to the own ship's heading (RELATIVE) or North (TRUE). When determining whether a close-quarter situation or risk of collision exists the use of relative vectors is the preferred setting. It is a good practice to switch between true and relative vectors to gain a better appreciation of the navigational situation. When using a true vector, own ship and other ships move at their true speed and course. True vectors can distinguish between moving and stationary targets. The relative vector helps to find ships on a collision course. A ship whose vector passes through the own ship's position is on a collision course. The Vector Length can be adjusted to the required time frame. It is useful to have both relative and true information visible simultaneously; this can be achieved by selecting relative vectors with true trails. Combining true vectors with true trails will give no indication of the relative movement of other vessels and the risk they present. To select the vector mode, shift the cursor to the vector mode box and left-click to select the vector required. The vector time can also be selected using the left button.

Last position

The LAST POSITION is a useful indicator. These history dots are placed at a fixed pre-set interval. Dots in a straight line at even spacing indicates a target travelling at a steady course and speed. Any changes in course or speed will become evident as the spacing becomes uneven. Change of course will not be shown in a straight line. A curve in the trail indicates an alteration of course whereas a change in the spacing of the plots indicates a change in the target's speed. These past data can also help the navigator to check whether a particular target has manoeuvred in the recent past, possibly whilst the navigator was away from the display on other bridge duties. However past positions, if used, can clutter the screen, and should be avoided in heavy traffic as the plots of different targets start crossing and overlapping each other and should be used with caution.

Marks

The MARK menu enables the navigator to mark any prominent targets or a point of particular interest. For example, the trackball can be used to select the desired mark from the mark box at the left-hand side of the screen. Also, you can drop the anchor mark by entering the Anchor coordinates provided by VTS in port areas in the MARK MENU. Right-click to open the mark menu and use the L/L to enter the coordinates.

Target tracking and AIS data box

The target tracking and AIS data box appears on the right-hand side of the radar screen. It provides information of automatically or manually acquired targets, including a display of the range, bearing, course, speed, CPA and TCPA, bearing course range (BCR) and bearing course target (BCT). The target list provides a comprehensive data display of all targets being tracked. To acquire a target on the radar screen, move the cursor to the target and left click. The TARGET ACQUIRE key on the keyboard can also be used to acquire the target. The CPA limit box can be used to set the range and time for CPA as required. If a target is predicted to breach the CPA limits, the alarm will sound and/or displayed.

Presentation modes

Radar users must clearly understand what they are seeing. North Up relative motion is the normal default radar display format. Within that setting relative and true vector and trails can be selected. The North Up mode shows the targets in their true (compass) directions from own ship, North being maintained up on the screen. The heading marker changes its direction according to the ship's heading. If the TRUE motion is used, own ship and other moving targets move according to their course and speed. Fixed targets such as landmasses appear as stationary echoes. The radar display provides the navigator with a bird's-eye view where other targets are portrayed relative to their own ship. It is an invaluable aid to navigation. Proper use and close monitoring of the radar, especially in reduced and restricted visibility, can help avoid close-quarter situations and/or collisions. It is therefore important that all radar operators understand its use and have a thorough knowledge of the equipment. As we might expect, the subject of marine radar is much more complex than has been set out in this chapter, which has only really skimmed the surface of the operation and use of marine radar systems. As a ship's navigator or OOW, it is crucial to be thoroughly versed in the principles and operation of marine radar and to study its faults and limitations.

In summary, marine radar technology has developed in leaps and bounds since the Second World War. Modern ship tracking systems are so advanced they can track and identify even small skiffs and power boats. This means that the ship's navigator can navigate more safely, and ships are able to detect hazards and manoeuvre out of harm's way earlier, reducing the potential for maritime incidents. In the next chapter, we will begin to explore in greater detail electronic chart and information display systems and how these have revolutionised modern-day seafaring.

6   Electronic chart and display information system

The Electronic Chart and Display Information System (or ECDIS) is a relatively recent development in the navigational chart system used on board naval vessels and merchant ships. With the use of the electronic chart system, it has become easier for the ship's bridge team to navigate and plot positions with greater confidence and accuracy. ECDIS complies with IMO regulation V/19 and V/27 of the SOLAS Convention as amended, by displaying selected information using a System Electronic Navigational Chart (SENC). ECDIS equipment that complies with SOLAS requirements can be used as an alternative to paper charts. Besides enhancing navigational safety, ECDIS greatly eases the navigator's workload with its automatic capabilities such as route planning, route monitoring, automatic ETA computation and Electronic Navigational Chart (ENC) updating. Furthermore, ECDIS provides many other sophisticated navigation and safety features, including continuous data recording which can be used for later analysis as well as the gyro, radar, ARPA, echo sounder, and so on.

The system can keep an accurate position as it uses GPS satellites which continuously feeds data to the onboard ECDIS system. Further enhancing its usefulness, ECDIS incorporates and displays information contained in many other nautical publications such as Tide Tables and Sailing Directions and integrates additional maritime information such as radar information, weather, ice conditions and automatic vessel identification. There are two main types of ECDIS charts currently available: the Raster Chart (RNC) and the Vector Chart (ENC). The RNC is a direct copy or scan of the paper chart. It looks identical to a paper chart and presents information exactly as it appears on the paper chart. The chart can be zoomed in or out or rotated. The ENC, on the other hand, is a computer-generated chart. The details on an ENC can be turned on and off depending on the requirement of the user. Objects on the ENC can be clicked for more detail. Depths can also be monitored to obtain grounding warnings. When zooming in, the features expand but the text remains unaltered.

The ECDIS system conforms to SOLAS requirements as well as several additional regulatory requirements, including SOLAS Chapter 5, Port State Control Requirements, IMO Performance Standards for ECDIS, S 52 Standard (Display Standard), S 57 Standard (Compilation Standard) and S 63 Standard (IHO Data Protection/Encryption Standard). ECDIS with adequate back-up arrangements may be accepted as complying with the up-to-date charts required by Regulation

DOI: 10.1201/9781003291534-6

V/20 of the 1974 SOLAS Convention. In addition to the general requirements for shipborne radio equipment forming part of the GMDSS and for electronics navigational aids contained in IMO resolution A.694 (17), ECDIS should meet the requirements of this performance standard. ECDIS should be capable of displaying all chart information necessary for safe and efficient navigation originated by, and distributed on the authority of, authorised hydrographic offices. ECDIS should facilitate simple and reliable updating of the electronic navigational chart. It should also reduce the navigational workload compared to using the paper chart. It should enable the OOW to execute in a convenient and timely manner all route planning, monitoring, and positioning currently performed on paper charts. It should be capable of continuously plotting the ship's position. ECDIS should have at least the same reliability and availability of presentation as the paper chart published by government hydrographic offices. Finally, ECDIS should provide appropriate alarms or indications with respect to the information displayed or any malfunction of the equipment. Updates to the ECDIS charts may reach the ship in various ways, depending on the capabilities of the service provider and the onboard communication facilities.

Benefits of ECDIS

The ECDIS system has several key benefits over the use of traditional paper charts, including:

(1) *Availability*. One of the great benefits of ECDIS over paper charts is the availability of electronic charts, especially when voyage orders are received at the last minute. Before electronic charts were developed, the second officer was responsible for sourcing the correct paper charts needed for a specific voyage by poring over the NP 131 chart catalogue. Once the relevant chart code had been found, the order would need to be placed with an onshore chandler. Hopefully, the chart would arrive in time; if not, the ship would inevitably be delayed until the chart arrived. Although not such a big issue for liner vessels, for ships operating on 'tramper trades' this presented an ongoing and costly problem. With the advent of electronic charts, the second officer's job is reduced to searching for the correct chart from an electronic database, selecting the correct chart, and downloading it direct to ECDIS. A task that once took hours, now takes just few a minutes.

(2) *Speed and accuracy*. With ECDIS as the primary source of navigation, the ship's navigator can plan and summarise the passage much faster than was the case with paper charts. Most ECDIS units have a facility where the waypoints can be imported into a Microsoft Excel® format, which reduces the amount of effort needed to manually input the waypoints when compiling the Passage Plan. Daily reporting data such as Distance to Go, Distance Covered, Average Speed, and so on can be computed quickly and with hardly any effort at all.

(3) *Corrections*. Before paperless navigation, the largest chunk of the ship's navigators time was consumed by correcting charts. Correcting charts with speed and accuracy was a skill that took a long time to master. Even then there was

a possibility of making erroneous corrections. The Temporary and Preliminary Notices to Mariners were especially tedious as these would predominantly come without tracings and required an ever-expanding file to be maintained. Indeed, keeping the world folio updated with the latest data was for many a navigating officer a great source of personal pride. Today, that has all changed with paperless navigation. The ship's navigator now receives weekly updates to the Electronic Charts as a zip file which is circulated by email and downloaded direct to ECDIS. Even Temporary and Preliminary Notices to Mariners are now sent directly to ECDIS.

(4) *Continuous monitoring of the vessel's position.* One of the single biggest advantages of ECDIS over paper charts is its ability to enable the navigator to see the vessel's position in real time without user intervention. ECDIS is interfaced with both the vessel's independent GPS transceivers, thereby providing redundancy even if one fails. That said, GPS is not infallible, and signals can be unreliable and prone to errors. This problem can be overcome by using the Radar Overlay and Echo Referencing facility available in ECDIS and radar. For this to function, the radars need to be interfaced with ECDIS. Once this is done, the navigator must activate the overlay tab on ECDIS which superimposes the radar screen over ECDIS. By checking that the radar echo matches with the ECDIS display, the navigator can be confident that the position displayed is indeed correct. A further feature is the continuous position monitoring, which is especially useful during coastal navigation. This uses ARPA Echo Referencing, a technique which is done by acquiring a fixed or stationary target such as a small island, lighthouse, or prominent rock on radar (ARPA) and then activating the ARPA tab on ECDIS. The next step is to deselect the Secondary Position Source on ECDIS and instead select Echo Reference. Once enabled, this gives the navigator a visual indication of the past tracks of both the Primary (GPS) and Secondary (Echo Reference) position-fixing modes. Finally, the navigator can also use the radar range and bearings to plot positions on the ECDIS display, just like on paper charts. All types of ECDIS come with an option of manually plotting the position using the range/bearing method. To do this, the navigator takes the range and bearing from a suitable radar object and plots this directly into ECDIS by using the range/bearing function on ECDIS itself. In ECDIS terminology, this is referred to as the Line of Position (LOP). A time stamp is printed on the ECDIS screen with both the GPS position(s) and the LOP. This serves as an indication of any offsets present between the GPS and radar fixes.

(5) *Anti-Grounding Alarms and Settings.* Though ECDIS has now evolved into a full-fledged primary source of navigation, it was initially intended as an anti-grounding aid to navigation. Today, the ability of ECDIS to warn navigators of approaching shallow waters makes it one of the most useful systems on the bridge. The flexibility of ECDIS means the navigator can determine what features they want to use, and which features they want to disable, albeit most SMS will have strict guidelines on the use of minimum safety parameter settings, such as:

(a) *Safety Frame (Look Ahead).* This is a setting which activates an alarm should the vessel exceed the specified safety limit. It should generally be set at not lower than 10 minutes in open waters depending on the speed of the vessel. This may be lowered in coastal waters based on the prevailing conditions.

(b) *Shallow Contour.* This setting indicates non-navigable areas and marks the boundary outside which the vessel may safely navigate. Crossing this boundary will likely result in the vessel running aground. The non-navigable area is usually indicated by a deep blue colour. To be effective, the shallow contour is usually set to the same value of the deepest draught of the vessel without any squat or ukc factored in.

(c) *Safety Depth.* This marks and highlights the minimum depth required for the vessel to remain safely afloat. As a rule of thumb, the safety depth is calculated as:

$$\text{Safety Depth} = \text{Deepest Static Draught} + \text{Anticipated Squat} + \text{Ship's} \times \text{Minimum} \times \text{UKC}$$

(b) *Safety Contour.* In general, the Safety Contour is set equal to, but not lower than, the Safety Depth setting. Waters with depths lower than the Safety Contour should be considered strictly no-go areas. The master may set the Safety Contour to a value higher than the Safety Depth if they determine that an additional safety buffer would be advantageous depending on the prevailing conditions. The Safety Contour is indicated by a grey coloured area.

(c) *Deep Contour.* This setting is useful for vessels engaged in operations such as tank cleaning or ballast water exchange where it is mandatory to carry out the operation in waters exceeding a certain depth. Vessels not engaged in these types of operations may set the value as appropriate but, in any event, this should not be lower than the Safety Contour. On ECDIS, the Deep Contour is indicated by a white coloured area.

(6) *ECDIS enhances the ship's Search and Rescue capability.* Modern ECDIS units have the added option of interfacing the ECDIS display with NAVTEX and EGC. Warnings and alerts are automatically displayed on the ECDIS screen, whilst at the same time sounding an audible and visual indication on the unit itself. Quick Range and Bearings are obtained by the Electronic Range and Bearing Line (ERBL) function. This enables the navigator to quickly determine if the vessel is in a position of aiding a distressed watercraft. The ECDIS unit also has a MOB function which can be activated in the event of a person falling overboard. This marks the position and datum which is used as a reference for rescue and recovery.

(7) *ECDIS is cost-efficient.* Although electronic charts are by no means cheap, they provide a more cost-effective alternative to paper charts. Electronic Chart Permits are obtained electronically direct from the supplier. Paper charts, though, must be physically delivered to the ship which involves postage and

agent handling fees. If the chart must be received at short notice or during out of office hours, these fees can be considerable. It was not uncommon before ECDIS for ships to alter course and divert to the nearest port just to pick up new charts. This, of course, entailed huge costs in fuel, agency fees, boat costs and so forth. By receiving the chart direct to ECDIS, these issues are completely avoided.

(8) *ECDIS is environmentally friendly.* In a world that is becoming increasingly conscious of the environment, it is natural that attention should turn to the vast waste created by disposing of old charts, many of which are never used. Add to this the endless stream of print out blocks, tracings, and Temporary and Preliminary Notices to Mariners, the environmental cost of paper navigation really cannot be ignored. Whilst there are environmental costs associated with ECDIS, these are minimal.

In summary, there are clear benefits and advantages to using ECDIS over traditional paper charts. Voyage information and navigational data are processed and displayed in real-time, passage planning is faster and easier to complete, alarms and indications are centred in one single location, chart corrections can be done with ease, and the chart can be tailored to the specific needs of the navigator and the passage. But it should be noted there are also several disadvantages that come with ECDIS.

Disadvantages of ECDIS

(1) *Over-reliance.* ECDIS is in some respects a victim of its own success. A highly capable and accurate resource, the use of ECDIS can lead navigators to over-rely on it as the single point of navigation. As is the case with over-reliance on anything, relying solely on ECDIS can lead to skill fade. Good navigation skills require constant practice. No matter how good the ECDIS system might be, its performance still relies on human input, management, and interpretation. A vessel that has switched its AIS off will not be visible on ECDIS. If the radar overlay is not turned on, the vessel will not be detected. This means visual watchkeeping is just as important in the age of electronic navigation as it was in the days of paper charts. There really is no substitute to manual navigation and watchkeeping techniques. If the ECDIS display fails, navigators must resort to traditional methods, including manual radar plotting, sights and calculating compass errors. With an increasing dependence on technology, it is vital these key skills are not forgotten.

(2) *Garbage In Garbage Out (GIGO).* ECDIS is, at the end of the day, an electronic system that depends solely on the inputs it receives. Erroneous position inputs from the GPS or a loss of GPS signal can have grave consequences, leading the ECDIS to go into dead reckon mode. If the alarm is missed, the result can be disastrous. Hence, it is critically important to check the performance of the sensors and to carry out frequent comparisons between the primary and secondary means of position fixing. Other inputs such as the gyro, anemometer,

echo sounder, and NAVTEX should be frequently and independently verified to ensure their efficient operation.

(3) *Wrong settings*. Feeding the ECDIS with wrong parameters for safety critical settings such as the Safety Depths and Safety Contours can lead to a false sense of security. It is extremely important that the master confirms these settings each time they are changed. These settings should be password protected and should verified at the start of each watch. Alarms should not be deactivated without good reason and never for the sake of avoiding frequent alarms.

(4) *Alarm deafness*. When alarms start sounding too frequently, the navigator may end up developing alarm deafness. This leads to the watchkeeper acknowledging the alarm without checking what the alarm is indicating. To avoid this, alarms should be carefully chosen, which are appropriate to the prevailing conditions. Any alarm settings not applicable to the immediate conditions should be disabled according to established procedures, recorded, and reset when appropriate.

(5) *System lag*. Modern ECDIS displays can show a vast amount of information at the same time. Combined with the myriad systems and equipment interfaced with ECDIS, the system can slow down leading to system lag. To avoid this from happening, it is important that the hardware which ECDIS uses is commensurate with the software installed. This may require upgrading the RAM and graphics card. Moreover, frequent upgrades are often necessary to keep the ECDIS working smoothly.

(6) *Different types of ECDIS software*. Learning to navigate with paper charts was a skill which had to be mastered only once. It was then just a matter of routine practice to stay proficient. With ECDIS, however, there is a vast array of systems and software available which means different ships will have different platforms installed. Whilst most companies aim to have similar – if not the same – systems installed across their fleets, it is not uncommon for navigators to move from one vessel to another which uses totally different navigational systems altogether. Whilst the basic functionality of ECDIS should be the same regardless of make or model, each platform will have its own unique nuances which the navigator will need to acquaint themselves with prior to assuming navigational watchkeeping duties. In answer to this, an increasing number of Flag states require navigators to undertake mandatory type-specific training before joining a new vessel. Logistically, this obviously provides considerable issues for shipping companies; therefore, it is becoming common practice for ship owners to replace disparate ECDIS systems with common framework systems such as TRANSAS, an ECDIS system owned by Wärtsilä.

(7) *Anomalies*. Navigators need to be conscious of the possible anomalies that may manifest themselves when using ECDIS. These can be as simple as the SCAMIN (Scale Minimum) function not being visible at certain scales.

(8) *Information overload*. When a system as advanced and capable as ECDIS is used, it is easy to fall into the habit of having as many settings showing at once. Whilst this may seem reasonable at the time, it can lead to informational

overload. This not only slows the system down (see point 5, System lag), but it also becomes much harder to discern relevant data from irrelevant dross. When using ECDIS, always try and keep the display as clean as possible; if using overlays, remove the overlay as soon as it is no longer required. Keep alarms and indicators to the minimum necessary for a given time or passage. Finally, it is always advisable to keep a hard copy of the passage plan just in case the ECDIS fails.

(9) *Resistance to change.* Although this may sound like a trivial issue, it can be quite problematic, especially when younger officers are working with more experienced officers and crew members. Although an increasing number of mariners today are taught how to use ECDIS during their initial training at college, there remains a sizeable cohort of older officers who have used traditional navigational techniques their entire careers. As with any professional setting, the introduction of new technologies and methods can be disconcerting. Transitioning experienced seafarers from paper-based navigation to electronic navigation requires retraining as well as tact and diplomacy.

ECDIS is an amazingly powerful tool and a vast improvement on paper charts. But its benefits are limited to the accuracy and ability of the people using it and the systems that feed it data. If AIS information is erroneous, then ECDIS will be erroneous or the ECDIS display might not be referenced to WGS 84 datum. Whatever the case, it is critically important that navigators retain currency in manual navigation techniques and avoid becoming overly dependent and reliant on ECDIS. The same of course applies to all electronic and virtual aids to navigation.

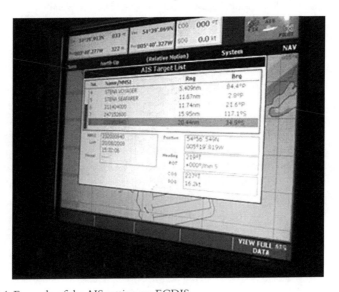

Figure 6.1 Example of the AIS setting on ECDIS.

Regulations and IMO performance standards for ECDIS

The primary function of ECDIS is to contribute to safe navigation. Subsequently, ECDIS must be provided with adequate redundant arrangements in the event of failure as per SOLAS regulation V/20. Furthermore, SOLAS mandates that ECDIS, in addition to all radio equipment forming part of GMDSS and electronic navigational aids, must comply with IMO resolution A.694 (17). This means that ECDIS must be capable of displaying all chart information necessary for safe and efficient navigation originated by and distributed on the authority of a government-authorised hydrographic office. Furthermore, ECDIS must be able to facilitate the simple and reliable updating of electronic navigational charts; in other words, ECDIS should not place a greater burden on navigational officers than would be expected if they used traditional paper charts. To ensure ECDIS meets this performance standard, the system should reduce the navigational workload of the navigator, including the execution of passage planning, monitoring, positioning, and plotting the ship's position. Furthermore, ECDIS must have the same level of reliability, availability, and accuracy as official paper charts. There are currently three main statutory frameworks that govern the use of ECDIS on board merchant ships. These are:

(1) SOLAS Chapter 5 – Safety of Navigation
(2) IMO Performance Standards for ECDIS, specifically S.52 Standard (Display), S.57 Standard (Compilation), and S.63 (IHO Data Protection and Encryption); and
(3) Port State Control Requirements which stipulate approved ECDIS types and back-up systems, type-approved hardware and software, approved installation, official chart services and the training and upkeep of navigational officer competency.

Types of ECDIS charts

There are two types of ECDIS charts:

(a) Raster Navigational Chart (RNC); and
(b) Vector Chart (ENC).

The raster chart is a direct copy or scan of the traditional paper charts. It looks exactly like the paper chart and has all the same information, markers, and diagrams. The only difference between the electronic and paper versions is the former can zoom in and zoom out and rotate. When the chart is enlarged or reduced, everything on screen grows or shrinks in size. The vector chart or Electronic Navigational Chart (ENC) is a completely computer-generated chart. The details and markers on the chart can be switched on and off as needed; objects can be clicked on for further information and depths can be continuously monitored. When zooming in or out, the features on screen expand or shrink but the text remains the same size.

Basic ECDIS settings and system requirements

Because the ECDIS system is a computer platform, there are specific settings and system requirements that are needed for the ECDIS to work. To simplify what is a complex piece of equipment, we can summarise the interface in terms of ENC layers, of which there are four.

(1) Display Base, where information cannot be deleted
(2) Standard Display
(3) Full Display; and
(4) Custom Display.

Irrespective of which display setting is used, there are three minimum interface requirements which cannot be disabled. These are:

(1) Position Sensor
(2) Heading Sensor; and
(3) Speed Sensor.

Apart from these core settings, the ENC is completely configurable to the user's own specifications.

Keeping ECDIS up to date

Updates to the ECDIS charts may reach the ship in various ways, depending on the vessel's service provider and the onboard communication facilities. There are four primary means of receiving ECDIS chart updates:

(1) On data distribution media (DVD)
(2) As an email attachment via SATCOM
(3) As a broadcast message over SATCOM; and (4) as an internet download.

Procuring ECDIS charts

As we mentioned earlier, the method of procuring electronic charts is different compared to paper charts. Unlike paper charts which are only purchased once – until a new edition is published, or a chart is cancelled – electronic charts are purchased through a licensing system called Permits. Permits are used to control the distribution and use of electronic charts and to prevent unauthorised copying and distribution. When a vessel requires a new chart, the master or navigating officer orders the correct chart via an online database, which replaces the NP 131 paper chart catalogue. The chart cells for the forthcoming voyage can be activated almost immediately by purchasing the relevant license which then activates the chart. Once the base DVDs are received, the charts can be uploaded directly to ECDIS. ENCs are issued by or on behalf of a national hydrographic authority and must comply with the specifications of the International Hydrographic Organisation

(IHO). To use the ENCs ships must contact an official ENC distributor such as Furuno® ECDIS.

The ordering cell permit is a three-step process. Let us suppose our ship will sail from Southampton to Sydney. We can use the Chartco Passage Manager to get the details of charts needed for the intended voyage. We do this by logging onto the Chartco Passage Manager database then: (Step 1) clicking on the main menu then selecting the option 'Routes and Passages'; (Step 2) selecting the automated routing function. Here we can input the ports of arrival and departure throughout our passage. The system will calculate the most expedient route based on each of the waypoints we entered earlier. This is only a provisional route intended to show which charts will be needed for that specific passage. The system provides extra functionality such as weather conditions, no-go areas, and known navigational hazards for relatively short passages and gyro compass track options for longer passages such as ours. The right-hand column under the ENC/AVCS displays all the charts needed. The user can then select the appropriate charts as required from each category. There are six different categories of ENCs:

(1) Overviews
(2) General
(3) Coastal
(4) Approach
(5) Harbours; and
(6) Berthing.

Each chart can be selected by clicking on the chart area on the map. The selected chart will have yellow lines and will display the cell name. As mentioned earlier, it is important to take weather conditions into account when selecting charts. For example, if transiting the Indian Ocean during the tsunami period, it may be advisable to consider more than route. The route from EDCIS can be exported and analysed using Met Manager, which is a weather-monitoring software platform. Also, during passage planning the waypoints can be adjusted accordingly. (Step 3) Once all the required charts are selected, we need to send a request to the Chart Distributor. This is done by selecting the option 'Add selected products to basket' at the right-hand bottom corner of the screen. A shopping basket icon at the bottom indicates that there are items within the basket. Once the charts have been added to the shopping basket, a second option – Create Order – will appear followed by an Order Address Information window. This is where the order information must be completed; for example, the name of the person ordering the charts and the billing address. Once the order has been successfully created, an Order Output Summary is generated where the order files are sent to the nominated chart agent. The system also prompts the user to create a PDF order form to be printed as proof of purchase. The order is then sent to an OUTPUT folder. From here, the user can open the folder and attach the order files to an email, which are sent to the nominated chart agent. Once the agent receives the order information, the agent will supply the AVCS ENC permit to the vessel as a zip file. This zip file can then be extracted to a USB stick, CD, or floppy disk. When the chart is needed, the media file is uploaded to ECDIS as a System Electronic Navigational Chart or SENC. Permits are usually issued for a period of 3, 6, 9 or 12 months.

Using ECDIS

Importing and downloading the permit

In this example, we will reference the Furuno® ECDIS. To load the permit into ECDIS, the following steps should be followed: (Step 1) Choose 'Chart Permits' from the Chart Menu. (Step 2) Click the 'Load File' button. The Load File button is used to initiate the ENC permit installation to ECDIS. When selected, a Permit Load window will display. In this window the user will have to identify the source, folder, or external drive where the ENC permits are stored. (Step 3) Select the correct drive from list box; usually drive 'E' is for USB sticks and drive 'A' is for floppy disks. (Step 4) The permit will be made available in two formats: one in PMT format and the other as a TXT file. The user must select the 'PERMIT.TXT' file then click to install the ENC permits. After importing the cell permits, the ENC will be automatically available in ECDIS. A backup of the permits can also be created by using the 'Backup' button in the selected drive. The format for backup files is always PMT. It is important to record the date of purchase and expiry date for each permit. Once the permit has expired, it is not possible to install the chart or to receive updates. If the chart has already been installed prior to the expiry date, the chart will still work but will not be updated with the most current data. A graphical coverage of all charts loaded onto the ECDIS hard drive can be viewed using the Chart Catalogue on the chart menu. The 'View Filter' section can be used to display charts for which permits are available. The charts appear colour coded. The easiest method to check if charts are up to date is to view the catalogue and check that the names of all required charts appear green. The names of charts which are not up to date appear orange.

Updating ECDIS

Charts on ECDIS need to be updated and corrected on a regular basis. Unlike paper charts where the procedure involves manual insertion of the corrections, correction to ENCs is made automatically. ECDIS allows updates to existing charts as well as replacing existing charts by installing new charts. Admiralty Vector Chart data is delivered on a series of CDs, which are labelled either Base or Update. They may also be supplied DVDs. New ENCs can be loaded with an ENC Base CD or DVD. This base CD is issued every six to eight weeks. The name of the countries included on the Base CD/DVD are written on top of the cover. Only those ENC Base CD/DVD are needed to be loaded for which permits are available. The Base disc label contains the date and week of issue. After all the relevant Base discs have been installed, ECDIS can now be updated. When using the AVCS DVD service only one disc needs to be installed as it contains the entire base ENCs and Updates up to and including the date of issue.

Charts are updated on a weekly basis and the ENC Update CD/DVD is issued every week thereafter. The only exception to this is in a week when the AVCS

Base disc is reissued. In these situations, an AVCS Update CD will not be issued. The Update CD contains all new edition ENCs and ENC updates issued since the last set of Base discs were issued. The ENC Update CD is cumulative, which means if more than one Update CD is received, only the latest disc need be used. In the case of DVDs, it is likely two DVDs will be received: one containing the Base and the other containing the Updates for that week. The procedure for loading the ENC Bases or Update CD is identical. This can be done from the Load and Update Charts function located on the Chart Menu:

(1) Insert the CD into the ECDIS CD-ROM
(2) Select CHART MENU
(3) Select LOAD AND UPDATE CHARTS
(4) Select FROM CDROM
(5) Select the location of the CD-ROM and click LOAD.

The charts will then begin to load into ECDIS. In addition to the automatic updates explained above, ECDIS also allows navigators to manually update the ECDIS system to apply their own navigational information as a supplement to the existing ENC data. This information remains separate from the contents of the ENC data already installed on ECDIS, however, it can be displayed in conjunction with the ENC data. Manual updating should normally be used when the same information cannot be added via the automatic updates. This might include where the latest navigational warnings are issued by radio, through NAVTEX or SAT C. Manual corrections can be deleted when they are no longer required. The method varies significantly between different ECDIS types. The Operators Manual should always be consulted before attempting to manually update ECDIS. Again, referring to the Furuno® ECDIS, the process for manually update is as follows:

(1) Under the Chart Menu, select MANUAL UPDATES. Select PLANNING within the MANUAL UPDATE dialogue box
(2) Select NEW then select the type of object to be added (for example, a buoy). Select ACCEPT NEW OBJECT. This will upload the new data directly into ECDIS.

Using the Admiralty Information Overlay (AIO)

The Admiralty Information Overlay (AIO) contains all Admiralty Temporary and Preliminary Notices to Mariners in force worldwide, including those that have been incorporated on to paper charts but have not yet been included in the ENCs. AIO makes passage planning simpler and safer by showing where important TPNM changes may impact on a passage. AIO can be used if the ECDIS is compatible to show AIO. Some countries still do not issue ENC TPNMs in AIO. This is because TPNMs have not yet been fully integrated into ENC data by all national hydrographic offices. Depending on the vessel's route and the ENCs used, the system should confirm whether or not onboard ENCs include TPNMs. If the onboard ENCs do not include TPNMs, then these should be manual corrected.

Setting the safety contour

ECDIS has become the essential tool for watchkeeping officers' onboard ships. Navigating a ship with ECDIS is fundamentally different from navigating with paper charts. It is therefore important that masters, navigating officers, and ship owners are aware of the benefits of proactively managing the ECDIS chart display, safety settings, and alarm systems. Sadly, ECDIS equipped ships have been involved in several groundings which could have been avoided had it not been for failures in the setup and use of ECDIS safety settings and alarm systems. Inappropriate settings are likely to render the safety contour alarms meaningless. The use of ECDIS safety settings has often been overlooked by navigating officers due to either insufficient training or complacency. Where ECDIS is newly installed on ships, navigating officers may be unfamiliar with the setup and use of ECDIS alarms, thereby increasing the risk of grounding in shallow waters. Appropriate safety settings are of paramount importance for using ECDIS. These settings control how the ECDIS presents depth information, making it easier to visualise areas of water that are safe for the vessel to navigate in from those which are not. For this example, we will continue to use the Furuno® ECDIS. Safety contour parameters can be set by following these steps:

(1)   Click on the main menu and select CHART DISPLAY
(2)   Select the MAIN TAB to display it.

As we have said previously, the safety contour is the most important parameter of all the safety settings as it displays unsafe water areas. Through its integration with the radar and echo sounder, the safety contour helps detect isolated dangers and triggers the anti-grounding alarm. In its simplest form, the safety contour is an outline which marks the division between safe and unsafe waters. The colour blue is used to indicate the unsafe areas whereas white and grey are used to signify safe areas. The default safety contour, if not specified by the navigator, is set to 30 metres (98 ft). The blue colour on a traditional paper chart does not provide a vivid picture of shallow water, i.e., the depths mentioned in the blue part of a paper chart may be shallower for a deep draft vessel whilst safe for a vessel with a higher draught. Unlike paper charts, ECDIS allows the navigator to set safety parameters according to the ship's static or dynamic particulars.

The safety contour is calculated by:

$$\text{Ship's Draught} + \text{Squat} + \text{UKC} - \text{Height of Tide}$$

To use an example, let us assume out ship's draught is 10 metres (32 ft). The ukc mandated by the SMS is 10%. Remember, the ukc calculation considers various factors such as sea conditions, salinity, and water density, and increases in draught caused by rolling. The ukc should always be calculated using the company's ukc formula. In this instance the ukc is one metre (3.2 ft). For simplicity, we can assume the squat at maximum speed is also one metre (3.2 ft) and the height of tide is one metre (3.2 ft). The safety contour would therefore be equal to 11 metres (36 ft):

$$10\,\text{m} + 1\,\text{m} + 1\,\text{m} - 1\,\text{m} = 11\,\text{m} \text{ or } 32\,\text{ft} + 3.2\,\text{ft} + 3.2\,\text{ft} - 3.2\,\text{ft} = 35.2\,\text{ft}$$

Safety contours are presented in values of 5, 10, 15, 20, 30 and so on. If the value set by the navigator is not available among the available depth contours, ECDIS selects the next deepest available contour in the ENC. If within a specified time set by the navigator, the ship is about to cross the safety contour, an alarm will sound. Based on the value of safety contour, ECDIS displays the isolated danger symbol for underwater features and obstructions which may pose a danger to navigation. The Isolated Danger Symbol is displayed if any underwater feature, such as wrecks, rocks, or other obstructions, have a depth less than the safety contour in waters beyond the safety contour.

Safety depth settings

The sole purpose of the safety depth is to portray spot soundings either in grey for deeper depths or in black for shallower depths. This compares to the safety depth value entered by the navigating officer, thereby highlighting potentially safe and unsafe areas. The safety depth value has no effect on alarms or any other aspect of ECDIS.

The safety depth is normally calculated as ship's draught plus squat. You may be wondering why we need to mention the safety depth when the safety contour shows us perfectly well where the safe and unsafe waters are? The answer to this is quite straightforward. It is perfectly logical to select the Safety Depth equal to the Safety Contour. Some soundings on the shallower side of the safety contour will be grey because they are deeper than the safety depth set by the navigator, although the depth may be deeper than the safety contour selected by ECDIS. The depths below the safety contour may not always be non-navigable. Suppose, for example, if the safety depth and safety contour are both set to 11 metres (36 ft) the ECDIS will emphasise the depth contour equal to or deeper than the selected contour, which, say, is 20 metres (65 ft), whichever is available in the ENC. Thus, we can see that water areas with depths between 11 metres (36 ft) and 20 metres (65 ft) are navigable but are below the limit set by the safety contour. This provides the navigator with additional information about where the ship could safely pass if crossing over the safety contour is required (an alarm will still sound). This could provide additional manoeuvring room in narrow passages where safe depths exist. There is also a possibility that depths shallower than the safety depth may exist at one point in the navigable waters. The safety depth setting will then highlight this hazard.

Zones of Confidence

When calculating safety depth, it is equally important to consider Categories of Zones of Confidence (CATZOC) features or Zones of Confidence (ZOC). It is not uncommon for the survey data displayed on ENCs to be some years old. This raises the question of how confident navigators should be when passage planning, especially in coastal areas and narrow channels. The ZOC is a tool which is used to determine the accuracy of the underlying hydrographic data. This information is available in six different categories. This enables the navigator to make educated assumptions regarding the degree of reliance of the information presented on the chart when planning a passage or executing navigation. To understand the ZOC better, we should first consider the various aspects used during bathymetric surveys to determine the category of ZOC. The first aspect concerns position accuracy. Position accuracy refers to the horizontal accuracy of a depth or feature. Quite self-explanatory, position accuracy gives the navigator a better understanding of the position at which the survey was made. The second aspect concerns depth accuracy. Whilst not in order of importance, for the purposes of understanding, we can say that the depth accuracy holds extreme importance to determining the ukc, thereby minimising the risk of grounding, bank and canal effect, squat, and so forth. Depth accuracy refers to the vertical accuracy of recorded depths. The third aspect is seafloor coverage. This provides us with an idea of the extent of the area that was used to obtain the seafloor data. Obviously, the wider the area covered, with a higher frequency of data taken, will help ascertain the reliability of the bathymetric data. And finally, survey characteristics. This section of the ZOC diagram provides specific information pertaining to the exact category of the data. For example, in a category A1 area (explained below) it may be stated that the data were obtained with the best possible survey techniques and accuracy using both DGPS and LOP. By combining this information, together with the ZOC diagram, and the details of the survey, including the date of the survey, we can categorise the survey data according to a its level of accuracy with A1 being the most accurate (Table 6.1).

To put this into some context, let us consider an example. Our ship has a draught of 7.7 metres (25 ft) and squat of one metre (3.2 ft). this means the effective draught is 8.7 metres (28.5 ft). The required ukc is 10% of the deepest draught, which is 0.87 or approximately 0.9 metres (3 ft). We see that the total safety depth required complying with ukc policy is 9.6 metres (31.4 ft). The safety depth value can be set at 10 metres (32 ft). However, we have not yet considered the depth accuracy as per the ZOC. To do this, let us surmise the CATZOC for this area is category B, which implies there can be an error of one metre + 2% of depth, which equals 1.2 metres (3.9 ft). Therefore, if CATZOC error is allowed, the minimum depth required for our vessel would be 10 metres + 1.2 metres equalling 11.2 metres (32 ft + 3.9 ft = 36.74 ft). As the safety depth cannot be entered into ECDIS as decimals, we must enter 12 metres (39.3 ft) as our safety depth. During passage planning, it is essential that the CATZOC is displayed and noted for all stages of the voyage.

Shallow contour

The shallow contour highlights the gradient of the seabed. It is the grounding depth, i.e., the depth below which the ship will run aground. This value can be set equal to the ship's draught. Therefore, if the ship's draft is 7.7 metres (25.26 ft) the shallow contour value can be set to eight metres (26 ft). ECDIS will then display the next depth contour available in the ENC. All areas between the zero-metre depth and the shallow contour are therefore not navigable at all and appear hatched. As we have already discussed earlier, the division between safe and unsafe water is highlighted by chart colouring, with a blue colour for indicating unsafe areas and white or grey indicating safe areas. The unsafe area is further defined with the selection of shallow contours showing dark blue in the shallow water and

Table 6.1 Description of zones of confidence

ZOC	Position Accuracy	Depth Accuracy		Seafloor Coverage	Typical Survey Characteristics
A1	± 5 m + 5% depth	=0.50 + 1%d		Full area search performed. Significant seafloor features detected, and depths measured	Controlled systematic survey high position and depth accuracy achieved using DGPS or a minimum three high quality lines of position and a multibeam, channel or mechanical sweep system
		Depth (m)	Accuracy (m)		
		10	± 0.6		
		30	± 0.8		
		100	±1.5		
		1,000	±10.5		
A2	± 20 m	= 1.00 + 2%d		Full area search performed. Significant seafloor features detected, and depths measured	Controlled systematic survey achieving position and depth accuracy less than ZOC A1 and using a modern survey echosounder and a sonar or mechanical sweep system
		Depth (m)	Accuracy (m)		
		10	± 1.2		
		30	± 1.6		
		100	± 3.0		
		1,000	± 21.0		
B	± 50 m	= 1.00 + 2%d		Full area searches not achieved; uncharted features, hazards to surface navigation are not expected but may exist	Controlled systematic survey achieving similar depth but lesser position accuracies than ZON A2 using modern survey echosounder but no sonar or mechanical sweep system
		Depth (m)	Accuracy (m)		
		10	± 1.2		
		30	± 1.6		
		100	± 3.0		
		1,000	± 21.0		
C	± 500 m	= 1.00 + 2%d		Full area searches not achieved; depth anomalies may be expected	Low accuracy survey or data collected on an opportunity basis such as soundings on passage
		Depth (m)	= 1.00 + 2%d		
		10	± 1.2		
		30	± 1.6		
		100	± 3.0		
		1,000	± 21.0		
D	Worse than ZOC C	Worse than ZOC C		Full search not achieved; large depth anomalies expected	Poor quality data or data that cannot be quality assessed due to lack of information
U	Unassessed – the quality of the bathymetric data has yet to be assessed				

light blue between the shallow water and the safety contour when the four-shade display is selected.

Water contour setting

This is normally set to twice the ship's draught. However, navigating officers can use deep water contour values any way they want. ECDIS also gives the option of simple two-colour shading. In this situation light blue and deep blue will merge into a single blue colour and grey and white will merge to a single white colour. If the value of the safety contour is changed, the boundary between two depth shades changes accordingly. Two depth shades can be used during night-time sailing, though with extra caution, as this reduces the contrast difference between the adjacent depth areas.

Daytime and night-time settings

Figure 6.2 shows a comparison between the two shade and four shade depth patterns in daytime and night-time.

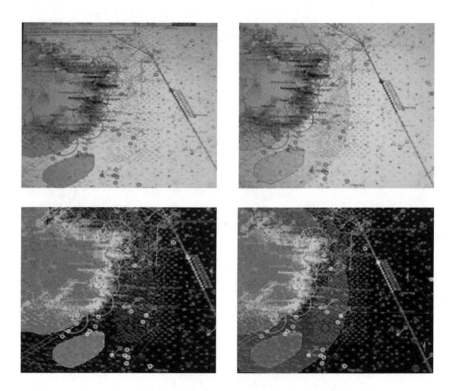

Figure 6.2 ECDIS shallow water contouring (daytime and night-time). Two-shade daytime, Four-shade daytime, Two-shade night-time, Two-shade night-ti.

Watch vector and anti-grounding function

The look ahead or watch vector compares the safety settings that have been entered by the navigating officer with the depth information contained in the ENC and generates an indication or warning where the safety settings will be contravened. It provides advance warning of dangers, primarily intended to prevent grounding. It acts as a final layer of safety should a navigational danger be missed by visual check or route scan. The scanned area is sometimes displayed as a cone or column on screen and should be set to a distance appropriate to the amount of navigable water ahead of the vessel. This value should be determined for each stage of the voyage and noted in the passage plan. Many officers fail to realise the significance of the safety contour and do not make proper use of the look-ahead vector. Using the Furuno® ECDIS as our example, this is the process for setting the watch vector:

(1) Go to the chart menu and select INITIAL SETTINGS
(2) Open the menu displayed on the left-hand side and choose ALERT PARAMETERS
(3) Click the CHECK AREA tab
(4) Set the AHEAD TIME or AHEAD DISTANCE
(5) The AROUND field allows the navigator to set fixed areas.

Note that the chart alert always uses the largest-scale chart available, which may not be the chart size the navigator chooses to use. The 'Chart Alert' feature should be highlighted to trigger the audible alarm whenever the safety contour is breached. It is a requirement to amend the alarm parameters from their previous settings when beginning a new voyage. The alarm parameters need adjusting throughout the voyage to ensure they are optimised for the prevailing conditions.

Hopefully by now it should be evident that ECDIS is a highly valuable asset which assists navigators, providing them with greater situational awareness than is available from manual navigation alone. However, unless ECDIS is used properly, it will be more of a hindrance to safe navigation than an aid. Nor can it be stressed enough that ECDIS is not designed, nor intended, to be a substitute for traditional navigational techniques and should be used only as an aid and not a replacement.

7 Ship handling in challenging conditions

Handling a ship in congested or high traffic areas is no easy task. Congested waters are high-density traffic areas where a vessel is likely to collide with another vessel if ship navigation is not carried out correctly. Open water navigation is not especially easy either, but by virtue of lesser traffic as well as the availability of sea room it does not pose as much of a threat as that posed by congested waters. Approaches such as those found in Singapore or China have dense traffic from fishing boats, small watercraft as well as vast numbers of large merchant vessels which, by default, equates to a higher degree of precaution, situational awareness, and decision-making.

Factors that make navigating congested waters difficult

Factors such as the state of the sea, weather conditions, the proximity of vessels, the status of the ship (underway or making way, laden or under ballast condition, etc.), interactions with other vessels causing differences in bow or stern pressures, proximity to land, the ship's steering gear effectiveness, the availability of aids to

Figure 7.1 Ships at anchorage, Indian Ocean.

DOI: 10.1201/9781003291534-7

navigation, rudder movement, and the effect of shallows are all factors which the navigator must take into consideration when passing through difficult and challenging waters. To ensure vessels can pass through waters such as these safely, it is imperative they adhere to the COLREGS without exception.

Duties of the OOW when navigating congested waters

When navigating through congested waters, the OOW must know their duties well and be fully aware of the positions of ships in the vicinity and of other fixed structures. Practical ship handling in these situations is a skill that is only acquired with experience. There is no shame in calling for the master if there is any doubt or concern about the safety of the vessel. After all, the master is responsible for the ship, its crew, and its cargo! To assist the OOW in their duties, there are several points which, if followed, will make their job considerably easier:

(1) Always comply with the COLREGS – they exist for good reason
(2) Be knowledgeable of the ship's manoeuvring capabilities and characteristics such as ukc, squat, stopping distances and so forth. If in doubt, consult with a senior member of the deck department or the master
(3) When entering a congested or difficult passage, change over from automatic steering to hand steering
(4) If appropriate, post additional lookouts. As per rule 5 of the COLREGS, "every vessel shall at all times maintain a proper look-out by sight and hearing as well as by all available means appropriate in the prevailing circumstances and conditions so as to make a full appraisal of the situation and of the risk of collision."
(5) Always keep the master informed about the current situation and course of action. If there is even the slightest doubt, call the master to the bridge
(6) Always try and give the engine control room sufficient notice regarding any manoeuvrability requirements; this requires good anticipation and skill in reading other ship movements
(7) Reduce the vessel's speed according to the prevailing conditions; it is better to arrive at port safe but late than on time and damaged. Refer to rule 6 of the COLREGS for further guidance on what is considered a safe speed
(8) Run parallel power units for the steering gears for additional safety. This ensures that should one of the steering gear systems fail, the vessel will not be left dead in the water
(9) Always test and confirm the working of the ship's whistle; like a car horn, the ship's whistle is an invaluable tool for navigating congested waters
(10) Adhere to the company's International Safety Management Code (ISM) checklist for navigating confined waters. The ISM will provide specific instructions and guidance for ship handling and responding to specific incidents
(11) Use all available resources including all aids to navigation and continuously monitor and cross-verify the charts. Use the largest-scale chart available for that area to check depths, distances, and bearings
(12) Coordinate with VTS as well as with the other ships in the vicinity using VHF; and finally
(13) Record all vessel movements in the ship's Bell Book.

Coastal passages are an extremely busy and stressful time for the ship's navigator and OOW. Traffic density is usually higher with an increased concentration of fishing boats and pleasure craft. This high-pressure environment demands expert time and resource management. When the OOW is occupied, they must ensure there is an alternative lookout on the bridge. The OOW must then spread themselves between position fixing and being the primary lookout; a multitasking skill if ever there was one! It is times like these that the OOW may feel they are overstretched and need assistance. If that is the case, the OOW should request the master to allocate additional resources to the bridge. Whilst this will not absolve the OOW of their watch duties and responsibilities, it should mean they are better able to complete more complex tasks, such as position fixing, whilst delegating anti-collision watch duties to junior members of the bridge team. When preparing to enter congested or difficult waters, there are several points that OOW is well advised to keep in mind: always:

(1) Follow the master's commands because

    (A) they have proven their ability, skill and have the experience; and
    (B) the master is ultimately responsible for the vessel

(2) Use the largest-scale chart available for increased accuracy
(3) Account for set and drift
(4) Identify navigational marks prior to approaching them; study the chart to identify any landmarks as early as possible
(5) Account for strong tidal currents that may cause deviation to the ship's course
(6) Comply with the enforcement rules of traffic separation schemes
(7) Use all available navigational aids to fix positions using any information available
(8) Adjust course to allow for deviations
(9) Avoid trading too close to navigational hazards such as oil fields, shallow areas, sandbanks and so forth
(10) Exercise extra vigilance in fishing grounds and waters shared with pleasure craft and cross-channel ferries
(11) Switch to hand steering when approaching shallow or congested waters
(12) Keep visibility on the ukc to avoid grounding and stranding
(13) Report to VTS as and when required
(14) Maintain constant communication with the engine control room and appraise with updates of progress
(15) Log all relevant information in the ship's logs, and, most importantly, the radio log; evidence of communication is critical in the event of an accident and subsequent investigation
(16) Increase the Plan Position Indicator (PPI) if the sea room becomes more confined
(17) Adjust course in the event the vessel starts to go off track; and
(18) Most importantly, call the master to the bridge before any critical situation arises. Attempt to assess and appraise the situation so you can advise the master when they arrive on the bridge.

Maritime chokepoints

A maritime chokepoint refers to a point of natural congestion along two navigable routes. Maritime chokepoints are generally caused by a high volume of traffic attempting to progress through a narrow passage. There are many maritime chokepoints around the world, including the Malacca Strait in the Indian Ocean, the Gulf of Hormuz in the Middle East, the Suez Canal, which links the Mediterranean Sea to the Red Sea, the Panama Canal, which links the Atlantic Ocean to the Pacific Ocean, the Strait of Bosporus (Turkish Strait), which links the Mediterranean Sea to the Black Sea; the three Danish Straits linking the Baltic Sea with the North Sea; and the Strait of Bab el-Mandeb, which forms the gateway for vessels passing through the Suez Canal to the eastern coast of Africa. Besides the inherent hazards associated with navigating extremely congested chokepoints, many of these locations have the unfortunate distinction of being subject to piracy and terrorism. It therefore goes without saying that extreme and constant vigilance is always needed.

Narrow channels and specific hazards for ships

There are separate rules dedicated to ship navigation with respect to narrow channels and fairways. A channel is defined as a natural or dredged body of water with shallow waters on either side, normally marked by buoys. A fairway is the same but in open waters wherein the depth of water on either side might not be shallow. The word narrow is highly subjective and is largely dependent on the characteristics of the vessel in relation to the characteristics of the water body. So why is there a rule dedicated to the navigation of such waters? Just as open waters present unique challenges for ships, narrow channels also present unique and specific hazards, some of which include:

(1) Prevailing or expected traffic volumes
(2) Currents and tides
(3) Bank effect and suction
(4) Turning points
(5) Vessel speed; and
(6) Tight passages.

Let us briefly examine these hazards in a little more detail.

Responding to prevailing traffic

The density of traffic is obviously going to be greater compared to that in open sea. With considerably reduced room available, the risk of collision increases exponentially. The OOW must therefore exercise due diligence during the transit, posting lookouts on the bridge wings as well as keeping an unobstructed watch on the radar. The OOW must ensure the vessel adheres to the COLREGS and anticipate the actions of other vessels who may – or may not – be as diligent in applying rule 9 of the Rules of the Road.

Bank effect and suction

When a ship moves through restricted waters, it must navigate close to the shore and other manmade structures, usually due to the limited navigable width of the channel. The shallow water and proximity of the sides of the channel has an adverse impact on how the ship behaves. These effects can cause errors in manoeuvring, which may lead to grounding or even collision. Importantly, any ship, regardless of their size, is heavily influenced by hydrodynamic effects when navigating through restricted waterways. The most common effects ships experience in these conditions include squat, bank, and bank cushion effects. Squat effect is caused when a ship moves through shallow water and some of the water that is displaced rushes under the vessel to rise again at the stern. This decreases upward pressure on the hull, making the ship sink deeper in the water than is normal and causing the vessel to slow. Squat effect increases with the speed of the vessel. Bank effect refers to the tendency of the ship's stern to swing towards the near bank when the ship is operating in a river or restricted waterway. Bank cushion effect occurs when the ship nears the bank and water is forced between the narrowing gap of the ship's bow and the bank. This water tends to pile up on the starboard side of the ship, causing the ship to sheer away from the bank.

Current

Current must always be accounted for when transiting narrow channels. Head currents (i.e., those flowing against the movement of the vessel) reduce the vessel's speed whilst retaining the ship's manoeuvrable speed. A following current (i.e., those flowing with the movement of the vessel), on the other hand, might cause a slight sluggishness in the ship's manoeuvrability but will otherwise not impact on the ship's transiting speed.

Effective communications

Effective bridge-to-VTS communications

When transiting through narrow channels it is vital to maintain communication with VTS. They are best positioned to provide guidance on how to proceed and how to respond to any of the issues discussed above.

Effective bridge-to-bridge communications

When leading, passing, or following other vessels, always maintain communications with the bridge using VHF. Navigational aids such as AIS and radar must also be used to determine the identity, positioning, CPA etc. of other vessels in proximity.

Effective bridge-to-engine control room communications

Prior to transiting narrow passages, it is important for the engine room to test the power and steering controls. This means advising the engine control room well in advance. Moreover, the engine control room must be advised of any specific requirements that might be necessary along the transit, such as heavy manoeuvring and quick response to changes in the propulsion speed and direction. Clear communication with the engine control room means they are prepared for and can respond to harder usage of the ship's engine machinery.

Accurate plotting when transiting narrow channels

It is critically important to constantly plot, at very short intervals, the position of the vessel on the chart during transit.

Turning points

The most challenging aspect of transiting a narrow channel is at the points of turning, which require a maximum application of skill and expertise. When on board, the pilot will be instrumental in ensuring the turning manoeuvre is performed safely. Turning in narrow channels requires an exact amount of rudder to be applied to achieve the correct phase at which the vessel should start turning, which is done in conjunction with the vessel's pivot point. A turn that is exercised too early is generally easier to recover from than one that is too late. Therefore, the helmsman should be supervised to ensure they do not apply more helm than is absolutely required.

High and low speeds

When transiting a channel, the vessel should always maintain a moderate speed appropriate for the width and depth of the channel. This provides a margin within which the vessel can operate across a larger range of speed. It also enables the vessel to increase speed to improve rudder responsiveness without going too fast. Lesser speeds call for the usage of bow thrusters, which may not be as effective as expected, especially compared to normal channel speeds. The engines must be ready for manoeuvring at the master's or pilot's command.

Using the bow thrusters

Handling a ship in currents, wind and tide is always difficult. These three factors can laterally shift the vessel from its course. The ship's pilot must keep an eye on these effects and constantly take remedial actions. Moreover, berthing in heavy winds and tides usually requires tugboat assistance, be it a towing tug or a checking tug, or a tug that imparts athwartship force to counter lateral drift. These are in addition to the ship's engines, rudder, and thrusters. Now, at any given time, the berthing speed is always kept on the lower side. It is best to maintain a speed of less than two knots under most circumstances. The bow or stern thrusters, which impart lateral or athwartship forces, are effective only when the speed is minimal. The thrusters work by dampening the impact of the vessel when it falls on the pier or berth. When the bow is falling heavily onto the jetty, the bow thruster is run to port or starboard, so that the fall is restricted, and lines can be passed safely with the heaving lines, jolly boat, or motorboat from the bow. When a ship is taken astern, which the ship's pilot normally does when approaching the jetty, the ship's stern goes either to port or to starboard. Subsequently, the bow will follow in the reverse direction, i.e., starboard or port. This effect is called canting. For a vessel with a right-handed propeller, i.e., for a propeller that rotates clockwise while the vessel goes ahead, the stern cants (swings) to port and the bow to starboard. The reverse happens for a left-handed propeller. At this stage the bow thruster comes into play. The bow, in the first instance, is given to port so that its swing to starboard is restricted while in the second case it is given to starboard.

You might be asking why the bow and stern swing when the ship comes astern? This is because of transverse thrust and screw race effect. The bow thrusters are driven by motors. The power of the bow thruster generally depends upon the power of the motor. An 800 to 1,000 bhp bow thruster is effective enough for a vessel with a LOA of 150–160 metres (492–524 ft) and a beam of 22–25 metres (72–82 ft) with a gross tonnage of about 10,000 to 12,000 tonnes. It can easily counter offshore winds of 3–4 Beaufort (11 knots) whilst berthing. In areas of higher wind speeds and ships having higher LOA and GRT, a more powerful bow thruster is required. Bow thrusters are generally installed to replace the use of tug-boats, which are hired by vessels coming into port. This reduces the operating costs of the vessel. In some ports, however, tugboat assistance is mandatory. In this case, the bow thrusters work in additional to the tugs. When anchoring, the bow thrust-ers also play an important role as they help keep the bow away from the anchor chain. When turning a vessel, such as at a congested dock, the bow thrusters help the vessel to spin on its axis. In these situations, the vessel is stopped and with the help of the bow thrusters the bow is swung to port or starboard, pivoting the stern. Occasionally, the stern is also rotated with the rudder and engines. But what is generally considered to happen in this instance is the vessel's natural tendency to cant whilst coming astern. If the vessel has a right-handed propeller, the vessel's bow will swing to starboard, while the engine is run astern. Therefore, it is always advisable to turn on the starboard wheel whilst coming astern as this enables the vessel to keep turning in the desired direction. When backing down, it is advisable to take a towing tug at the stern and use the bow thruster only as and when needed. The tug straightens the stern as it pulls, and any effects of transverse thrust are thus neutralised. The bow thruster thus plays an important role in the manoeu-vring of ships.

Restricted visibility

Main actions when transiting through limited visibility

One of the most important duties of the ship's OOW is the safe and smooth navi-gation of the vessel. During its voyage, a ship must sail through different weather and tidal conditions. It is the responsibility of the navigating officer to know and understand the ship's sailing route well in advance and to prepare for all possible eventualities. When transiting a passage with limited visibility, there are key points the OOW should note.

These include:

(1) *Knowing the ship inside-out.* An effective navigating officer must know every aspect of their ship to prevent accidents and collisions. From the dimensions to the characteristics of the ship, the navigating officer should know how the ship will behave under various circumstances. For restricted visibility situations, it is particularly important the OOW knows the stopping distance of the ship at any rpm to control the ship during an emergency stop

(2) *Inform the master.* During restricted visibility, it is important that the master is present on the bridge. The OOW must call or inform the master regarding the navigating condition. The navigating officer should also inform the engine control room and request the duty engineer to man the engine room in the event the engine room is on unmanned mode

(3) *Appoint adequate manpower.* It is important that sufficient manpower is present on the bridge to keep a close watch on the ship's course. Additional personnel must be appointed as "lookout" at different locations on the ship. This typically includes on the bridge wings and at the bow. If there is traffic in the area, the navigating officer must inform the engine room to provide enough manpower so that the engine is also ready for immediate manoeuvring

(4) *Keep the foghorn ready.* When navigating with restricted visibility, always ensure the foghorn is working properly. If the horn is air operated, drain the line prior to opening the air to the horn

(5) *Reduce speed.* Reduce the speed of the vessel accounting for the visibility level. If visibility is severely reduced, bring the ship down to manoeuvring rpm

(6) *Ensure the navigation equipment and lights are working properly.* Ensure that all navigating equipment and navigation lights are working properly during periods of restricted visibility. The OOW must ensure that the navigation charts are properly checked for correct routing and that a good radar watch is carried out

(7) *Stop all other works.* Though it may seem obvious, never multitask during restricted visibility, even if there are more than sufficient people present on the bridge. Furthermore, cease all deck work and order the crew to stand easy in their cabins. This is to prevent injury to personnel working on open deck and ensures the crew can make way to the muster point in the event a collision or grounding occurs

(8) *Open and close the bridge wing doors.* Ensure that the bridge wing doors are kept open and unobstructed. If the ship is passing through a sand or dust storm, secure the bridge wing doors and close all windows

(9) *Shut ventilation.* If the ship is passing through a sandstorm, the ventilation fans and accommodation and or engine room ports must be closed to avoid sand particles from entering bridge, the accommodation block, and the engine room

(10) *Follow all procedures.* Follow all procedures pertaining to restricted visibility as per rule 19 of the COLREGS. Moreover, monitor channel 16 on the radio and ensure that all important parameters of the ship such as latitude and longitude, time, speed, etc., are noted in the ship's logbook.

Among the most dangerous conditions to navigate a ship in are: restricted visibilities caused by fog, heavy rain, or dust storms. When the ship's navigating officer receives information regarding such weather conditions, they should take all necessary precautions to ensure that the ship passes through the restricted visibility area without causing a collision or grounding incident.

Heavy weather and rough seas

Every seafarer wishes for smooth sailing and calm seas. No one likes to be at sea during stormy weather, albeit being an inevitable part of a mariner's life. Some of the most common forms of heavy weather and rough seas are caused by tropical depressions or storms, typhoons, cyclones, and hurricanes. These are generated by varying atmospheric pressures over different parts of the Earth. The Beaufort wind scale, which is used to classify wind speeds, classifies strong winds as near gale, gale, strong gale, storm, violent storm, and hurricane based on an ascending magnitude of wind force. Movement of the sun causes pressure belts to shift thus creating varying temperatures over land masses and water bodies, which in turn cause pressure differences. Tropical depressions occur most often in the middle latitudes and tropical cyclones tend to originate in the Intertropical Convergence Zone (ICZ). A depression may develop and travel in any direction whereas tropical storms are mostly found to follow predicted paths in both hemispheres. Tropical storms then recurve after following a particular track. It is therefore important for the navigating officer to predict the location, magnitude, and path of any storms which they may encounter during their passage. Sometimes, the force of an anticipated storm is so great it is best to avoid these regions altogether or to navigate with extreme caution.

To ensure the vessel can navigate as smoothly as possible through tropical storms, there are a few points which the OOW is best advised to follow.

(1) *Wherever possible, use available information.* Tropical storms and depressions are formed by pressure and temperature variations. Seafarers have access to information regarding seasonal areas and frequencies of occurrence through MSI via EGC, Admiralty Sailing Directions, Ocean passages of the World and several other renowned publications.

(2) *Study weather reports.* Often weather reports and weather telexes provide warnings well in advance about unsettled weather conditions. A careful study of navareas and weather reports can be instrumental in obtaining early warning about pending and developing storms. Frequent observations from the various meteorological instruments on board can also be used to confirm the accuracy of weather reports.

(3) *Keep away from the centre of storm.* Once the presence of a storm or depression is confirmed, it is vital to establish the distance of the vessel from the outer region of the storm, the location of the eye of the storm, the centre of the depression, and storm's track and path. Buys Ballot's law advises the face of the wind and centre of low pressure will be from 90 degrees to 135 degrees to the right in the northern hemisphere and the same on the left in the southern hemisphere.

It is advisable to keep at least 250 miles (400 km) away from the centre of a storm though some shipping companies may prescribe specific distances to be maintained in their SMS.

(4) *Check the stability of the vessel.* Always maintain regular checks on the stability condition of the vessel and its compliance with intact stability criteria. Damage stability conditions should be evaluated carefully before the start of the voyage as it will assure compliance with damage stability requirements. By doing so, the vessel can take on heavy weather ballast before or whilst proceeding through rough weather areas. Heavy weather ballast provides additional stability to the vessel and, by lowering the ship's centre of gravity, makes the vessel more stable as the centre of gravity increases. Heavy weather ballast tanks are specifically designated; if those tanks previously carried oil or bunker fuel, they must be crude oil washed before flooding with heavy weather ballast.

(5) *Use the ballast tanks to minimise free surface effect.* All ballast tanks which are slack can be pressed into service to minimise free surface effect. This also helps to increase the ship's centre of gravity. Furthermore, the well-planned stowage of cargo, ballast or both can also help minimise the number of slack or partially filled tanks.

(6) *Exercise caution when changing speed, angle, and direction.* Often the waves associated with a storm or depression causes a reduction in intact stability of vessels, which can cause the capsize or rolling of the vessel to very large angles. IMO circular MSC 1228 provides guidelines with respect to the careful reduction of speed, changing angle, and direction of encounter and adjusting encounter period of waves to avoid parametric or synchronous rolling motions.

(7) *Secure loose equipment/cargo on deck.* For vessels with a lesser freeboard, the decks are washed frequently by waves with greater magnitudes. Thus, the securing of loose equipment on deck with additional lashings must be undertaken to strengthen and prevent their loss at sea. Safety lifelines can be rigged on vessels carrying cargoes on deck. Additional lashing must be taken to secure anchors, lifeboats, life buoys and life rafts.

(8) *Secure weather and watertight openings.* Various weathertight and watertight openings such as side scuttles, hatch covers, portholes, doors, and manholes must be secured and closed to prevent any ingress of water. Leaking or damaged gaskets or inadequate securing of the covers for such openings may affect the integrity of the compartment they are protecting. Alarms and indicators for the closing of remote watertight doors and openings are provided on the bridge. Their operational state must be confirmed prior to the vessel departing port.

(9) *Secure doors forward of the collision bulkhead.* Special emphasis is provided to secure the doors and openings forward of the collision bulkhead. These might include the forepeak store and hatches, vents, and openings forward. These spaces often house forward mooring equipment and associated electrical or hydraulic machinery. Spurling pipe covers need to be cemented well in advance. Bilge alarms in such remote compartments should be tried and tested regularly to give early warning of any ingress of water or flooding. Any openings in the subdivisions of watertight compartments which can cause progressive flooding must be secured.

(10) *Drains and scuppers must be free.* All drains on deck and scuppers used for drainage of green water must be free to prevent any accumulation of green water on deck. Secure aerials and antennas. Antennas, aerials, stay wire clamps and lashings must be inspected before wind speeds accelerate. Winds of gale force and above can easily break and blow away aerials. Moreover, storms are often associated with lightning therefore all aerials and antennas must be earthed, and any low insulation alarms investigated.

(11) *Keep check on the rpm to avoid load fluctuations on the main engine.* Due to the unsettled movement of the vessel, it is common for load fluctuations on the main engine to occur. Careful setting of rpm can help keep main engine fluctuations within permissible limits. This should reduce the risk of the main engines sustaining damage.

(12) *Inform all departments.* All departments onboard – i.e., the deck department, engine room and galley – should be informed well in advance of any storm warnings so that all deck, engine and galley stores, hospitals, sick bays, and work areas can be lashed and secured. Any major overhaul jobs, working aloft or the lifting of heavy machinery on deck and in the engine room using overhead, or deck cranes must be postponed.

(13) *Finally, the morale of the crew should be kept high.* It is not uncommon for crew morale to drop during storms and depressions. Seasickness, nausea, reduced appetite, and poor sleep can all manifest themselves in the crew leading to poor judgements, frayed tempers, and anti-social behaviour. To avoid these, it is important to keep crew morale as high as possible.

Parametric rolling

Rolling and pitching is an inevitable occurrence that affects every ship at sea. Whereas rolling and pitching is very natural for most types of ships, a different phenomenon called parametric rolling is unique and very dangerous and seems to only effect container ships. The size of container ships has increased dramatically since the 1980s as companies look to maximise the volume of cargo they can carry on each ship. In 2006 Maersk launched what was then the world's first mega container ship – the Maersk Triple-E class. These mammoth ships have a carrying capacity of over 15,000 TEU containers, though by 2021, even these had been surpassed by a new generation of mega ships capable of transporting in excess of 23,000 TEU containers at any one time. The problem with such large vessels lies in the design of the bows. A large bow flare and wide beam decreases the frictional resistance generated when the ship's fore-end passes through water. This makes it more streamlined with the hull, which reduces drag, and increases the ship's speed. As the wave crest travels along the hull, it results in the flare immersion of the wave crest, causing the bow to drop down. The stability of the vessel varies as a result of the pitching and rolling of the ship. The combination of buoyancy and wave excitation forces pushes the ship from side to side.

A similar action takes place as the bow goes down in the next wave cycle, resulting in synchronous motion that leads to heavy rolling of up to 30 degrees in just a few cycles. This rolling is referred to as parametric rolling. This phenomenon occurs only when the sea condition is at the head/stern or within proximity. There are two pitch cycles: maximum and minimum. The period of the roll is half the natural rolling period which coincides with a large phase angle. The maximum roll always occurs when the ship is pitching down, i.e., when the bow is downward.

Effects of parametric rolling

Parametric rolling is an extremely dangerous condition as the ship can develop a roll of as much as 30 degrees within a few wave cycles. If not corrected, this can cause the ship to immerse on one side leading to deck flooding and potential capsize. Besides this worst-case scenario, there are number of other impacts parametric rolling has, including:

(1) Causing heavy stresses on the ship's structure, especially in fore and aft sections
(2) Causing extreme stresses on containers and their securing systems, resulting in failure and loss of cargo
(3) Parametric rolling is very unpleasant for the crew which can lead to service degradation
(4) Variations in the load of the ship's engine can lead to substantial and lasting damage.

What to do in the event of parametric rolling

If parametric rolling starts, the most important thing to remember is not to panic. Always try to keep calm and collected. If rolling and pitching occurs simultaneously, aim to avoid a head-on at sea, and change the ship's course. Always maintain a correct centre of gravity. The ship should never be too tender or too stiff. If available, deploy the ship's roll damping measures immediately.

Tropical Revolving Storms (TRS)

All seafarers are familiar with Tropical Revolving Storms (or TRS) – an intense rotating depression (a region of low pressure at the surface) which develops over the tropics. They consist of a rotating mass of warm and humid air which create thunderstorms with strong winds, flooding rain, high waves, and surge. Conventional forces are involved, normally stretching from the sea surface of the depression up to the tropopause. Some of the most important characteristics of a TRS that mariners ought to know are:

(1) They appear smaller in size than standard temperate depressions
(2) They form near the Intertropical Convergence Zone, which is a zone of known instability

(3) They have nearly circular isobars
(4) There is an absence of fronts
(5) They result in a very steep pressure gradient; and
(6) They tend exhibit great intensity.

Detecting the emergence of a TRS can be done through several different methods.

(1) *Maritime warning and maritime alert messages.* Radio, telex and NAVTEX should be set at the right frequencies for receiving maritime warnings and alerts. These should be monitored closely. Refer to the respective ALRS volume for more data and frequencies of radio stations in the vicinity. Telex, although rarely used these days, is also a very important tool that provides high accuracy. Not all storms may be detected by coastal meteorological stations, in which case, the shipboard equipment and crew observations are key
(2) *Swell.* When there is no sight of land, the TRS will likely generate increasing swells, providing an early warning. Normally, the swell approaches from the direction of the storm
(3) *Atmospheric pressure.* Changes in atmospheric pressure are usually detectable without needing sophisticated equipment. Where a change in atmospheric pressure is suspected, monitor the barometer. If the barometer reading falls below 3 mb or more for the mean reading for that time of year (refer to the Sailing Directions for accurate information of pressure readings) it is likely a TRS is developing. Remember that the barometer must be corrected for the latitude, height, and temperature to achieve an accurate reading
(4) *Wind.* Wind direction and speed is generally constant in the tropics. Any variation from the normal direction for the area and season, and any unusual increase in wind speed, are indications of an approaching TRS
(5) *Clouds.* Cirrus clouds may be visible in the sky. As the vessel gets closer to the vicinity of a TRS (300 to 600 miles or 482–965 km) the clouds will progressively lower and cover a larger area (altostratus). These are then generally followed by cumulus clouds as the TRS gets closer still
(6) *Visibility.* Although it might sound like an oxymoron, it is common to get exceptionally good visibility when a TRS is developing
(7) *Radar.* The radar can provide a fair warning of a TRS at about 100 miles (321 km). The eye of the storm may even be seen on screen. An area of rain surrounds the eye, causing clutter on the radar screen. Remember although the signs might be visible on the radar, by the time the ship is close enough to pick up the clutter, the vessel is probably already experiencing high seas, gale force winds and rough weather.

Course of actions to be taken in case of a storm

Although it is unlikely a vessel will choose to sail into a storm with all navigational aids and communication systems operating, sometimes it is unavoidable. In these situations, keep at least 50 miles (80 km) from the centre of the storm. If possible, it is best to be at least 200 miles (321 km) off the centre. Make good speed. A vessel speeding in the vicinity of 20 knots (23 mph or 37 kph), following a course taking her away from the eye, can easily outrun an approaching TRS. Of course, this should be done before the wind increases to the point that the ship's movement is restricted. As mentioned earlier, a swift fall in pressure typically indicates an impending TRS. The vessel should continue her course unless the barometer reading falls by 5 mb or, by 3 mb when accompanied by high-force winds. If the vessel is trailing the storm (i.e., is behind the storm) and within the navigable semicircle, there should be sufficient time and sea room to shift away from the centre.

Northern hemisphere

In the event the wind is veering, the vessel is likely to be in the dangerous semicircle. The vessel should proceed with maximum speed keeping the wind at 10 degrees to 45 degrees on the starboard bow (depending on speed). The vessel should turn to starboard as the wind veers. In the event the wind direction is steady or backs, such that the vessel is in the navigable semicircle, the wind must be brought well on the starboard quarter and the vessel should proceed with maximum speed. Turn to port as the wind veers.

Southern hemisphere

In the event the wind is backing, the vessel is likely to be in the dangerous semicircle. The vessel should proceed with maximum speed keeping the wind 10 degrees to 45 degrees on the port bow (depending on speed). The ship should turn to port as the wind backs. In the event the wind direction is steady or backs, such that the vessel is in the navigable semicircle, the wind should be brought well on the port quarter and the vessel should proceed with maximum speed. Turn to starboard as the wind backs.

In port

If the vessel is in port and a TRS is approaching, it is best to put out to sea. Staying put at the berth, especially with other vessels in proximity, can be highly dangerous. Even with the best mooring practices, it is doubtful the ship will remain secured.

Icepacks and icy waters

Navigating in icy waters can be a real task for ships, as the later moves cracking and smashing through the frozen and frigid seas. When making way towards areas with sub-zero temperatures and ice-covered waters, the master must exercise extreme caution and pay utmost attention to the type of ice, its thickness, and its exact location in the sub-zero navigation areas. Icy waters present a major constraint for ships with sea ice being on average two to three metres (6–9 ft) thick. Often this ice can only be broken up by a specially designed ice-strengthened vessels or icebreakers with an appropriate ice-class hull. Most merchant ships and fishing vessels which are not ice-strengthened must keep away from ice waters and sub-temperate areas. In many places, where the concentration of ice is at a maximum and the ice pressure is highest, even the most powerful icebreakers have difficulty forcing their way through. To avoid getting caught in an ice drift, ships often need an ice pilot and or an ice breaker to accompany the vessel.

Manoeuvring in ice

First and foremost, it is imperative that if any alternative route is available for the ship, ice water should be avoided at all costs. However, if ice navigation is inevitable, it should be made at right angles to the leeward edge where the ice is loose or broken. While manoeuvring through ice – if a floe – cannot be avoided then it should be hit squarely with the stem. Note that a glancing blow may damage the ship's shell plating or throw the vessel off-course, causing another unavoidable blow. Entry into ice should always be done at low speeds to avoid engine damage. Once into the pack, the vessel's speed can be increased to maintain headway. This prevents the ice floe from closing in on the hull, the rudder, and the propellers. If the ship is stopped by heavy concentrations of ice the rudder should be put amidships and the engines should be kept turning slowly ahead. This will wash away the ice accumulates astern and will help the vessel to fall back. In a close pack during ice navigation, always avoid sharp alterations of course and keep the speed sufficient for steerage way. Full rudder movements should be avoided or used only in emergencies.

Figure 7.2 Heavy lift ship *Bigroll Barentz* navigating through drift ice.

Posting a lookout

Always keep a vigilant lookout for leads (navigable channels within an ice field). Additional lookouts should be posted forward and at height for safety. Conning should be carried out from the ship's bridge to get a better view of the ice accumulation. Always keep in mind that the stern must be observed for the rudder's movement to avoid a flow from moving the stern towards it. In these cases, it is advised to post lookouts right aft with torches, whistles, and radios, to ensure the bridge is informed immediately of any danger to the propeller. This is extremely important for twin screw vessels. Reduce the ship's speed if ice passes under the hull.

Caring for the engine

During ice navigation, the engines should be kept continuously running and under manoeuvring conditions in such a way that ahead and astern movements can be easily carried out without delay. Similarly, engine movements from ahead to astern and vice versa should be made cautiously to avoid stressing the engine mechanisms in low temperatures. Furthermore, when ice approaches the stern when manoeuvring, short bursts of power from the engines should be made to prevent ice from accumulating.

Navigating through ice at night

Insofar as is possible, avoid navigating through ice at night. It is always preferrable to "heave to" since the leads or lanes cannot be seen. Most ice navigators stop the vessel Along the edge of the ice and leave the vessel drifting along with the pack. At night, seawater lubricated tail end shafts are in danger freezing. To avoid this, vessels with single screws should have their aft peak tank filled with water and maintain it at a warm temperature by means of steam hose injection. The vessel should keep her engines running with the propeller on low rpm to avoid seizing up with ice.

Anchoring in ice

Anchoring in heavy concentrations of ice should always be avoided; if the ice is moving then its force may break the cable. When conditions permit, anchoring can be carried out though it must be done in light brash ice, rotten ice, or widely scattered floes with the main engine on immediate notice. The anchor should be brought in as soon as the wind threatens to move ice onto the vessel. Even with the advent of new techniques and technologies for ice navigation, such as radar sensor images through cloud cover, infra-red images, and satellite images for a larger view of the surroundings around the vessel, it is vital to understand that ship's operations of any sort under the influence of sea ice are not only dangerous but also life threatening, and utmost care must be taken while navigating through ice water areas.

Navigational warnings

Information concerning the navigational safety of ships was often difficult to obtain in in the earlier days of maritime navigation due to the limitations of maritime communication equipment. This situation has improved greatly over the years.

As technology has advanced, the quality of shipboard communication equipment has improved, providing navigational warnings well in advance. The Worldwide Navigational Warning Service (WWNWS), which was established in 1977 for the promulgation of information on worldwide hazards to navigation for international shipping, provides warnings and information of incidents which may constitute a danger to navigation. Many of these navigational warnings are of a temporary nature, whereas others remain in force for several weeks and may even be succeeded by a Notices To Mariners.

As a navigating officer the first duty and responsibility are always safe navigation. A good navigator plans each passage carefully. But even the best-planned passages are subject to unforeseen changes. On receipt of navigational warnings or any matter relating to the safety of life at sea, the navigating officer should immediately check if there is any impact on the intended passage. If there is an impact, these warnings should be plotted on the appropriate electronic or paper charts. Navigational warnings are one of the most important components of chart correction. There are various types of navigational warnings, but the main types are:

(1) Casualties to large navigational buoys, primary lights, or lightships
(2) Changes in depth of water
(3) Dangerous wrecks or obstructions to navigation
(4) Floating dangers such as containers adrift, drifting buoys, and icebergs
(5) Newly established aids to navigation
(6) Gunnery or live fire exercise areas and underwater military activities
(7) Vessels aground in approaching port areas
(8) Any changes to existing or developments of new buoyage systems; and
(9) Gale warnings.

Before the advent of electronic mail and other real-time services, it was not uncommon for weekly notices to arrive very late onboard meaning chart corrections would be completed late. What ensured the safety of the vessel were navigational warnings. Plotting the navigational warnings on charts ensured the vessel was moving along a safe passage. Fortunately, today, navigational warnings can be received through NAVTEX and Inmarsat C, which is configured as an EGC receiver. NAVTEX is a GMDSS requirement for the reception of MSI in coastal and local waters. The Inmarsat C Safety Net is an internationally adopted, automated satellite system for promulgating navigational warnings and other safety-related information. Software platforms such as Chartco and websites like Navarea Coordinator can also provide EGC messages. Care should be exercised when taking navigational warnings from the internet. Never use data obtained from unofficial websites. Information regarding official and recognised websites can be received from ALRS Volume 5 under the section MSI.

The WWNWS divides the world into 21 navareas, which are identified by roman numerals. Any ship can sail in any one of these 21 navareas. To obtain navigational warnings from any of the 21 navareas, the area needs to be selected in the Inmarsat C. The best practice is to select all navareas that will fall under the passage plan at the start of the voyage. The OOW should ensure that the Inmarsat C EGC receivers are configured to receive MSI messages appropriate to their intended passage.

Navigational warnings from Inmarsat C

The EGC setup screen allows the navigating officer to choose the right navareas for which to receive messages.

Navigational warnings from Chartco

Navigational warnings can also be sought from Chartco. Navigational warnings can either be accessed from the navarea warning bar on the homepage or from the NAVAREAMANAGER button in the main menu. On selecting navarea warnings, three screens will appear, NEW, INFORCE or CANCELLED, depending on what outstanding actions are left. The setup option at the right-hand bottom of the screen allows the officer to set the navareas that they wish to collect and display warnings for. By default, all navareas are enabled and therefore updates for all 21 navareas will be collected unless the settings are adjusted accordingly. If the vessel does not trade in any of the default navareas, these can be deselected. Click on SAVE to enable the changes. After selecting warnings in the relevant charts, the user must click on APPLY. The warnings are then removed from the NEW view. On selecting a particular navigational warning message, the text is displayed on the right-hand side window. The PRINT SELECTED option is used to print a hard copy of the message. A list of in-force navarea warnings can be obtained from the REPORTS option. In-force lists of warnings can be received from SAT C every week. Navarea warnings can be shown on the map. Warnings are colour-coded to indicate their status. New warnings are coloured red, in-force warnings are amber and cancelled warnings are blue.

The first thing the OOW must do on receiving a navarea warning is to check if it will affect the current passage. If the warning falls within the current course, it should be plotted immediately and brought to the attention of the master and the other watchkeeping officers. If the warning is out of route, it should be signed by the OOW and filed. When plotting a navarea warning, care should be taken that the minimum information required is displayed without obscuring other chart data. On ECDIS, a USER CHART can be used to plot navarea warnings. It can be named after the warning number. On cancellation of the navarea warning, the user chart must be deleted. NOTES can be used to display information without cluttering the chart area. A systematic procedure should be adopted when responding to navarea warnings. Navarea warnings should be handled in such a manner that every officer is able to find out with ease which warning messages are plotted and what charts they affect. A good way to manage navigational warnings is to maintain a file divided into three sections: IN ROUTE, OUT OF ROUTE, and CANCELLED. A printout of the warning message should be taken out and categorised based on the above-mentioned sections. If the warning affects the intended route, the affected paper charts if used in addition to ENCs must be annotated on the printout. The charts must then be plotted with the warning as well as the warning number. Navarea warnings are typically cancelled for one of three reasons:

(1) The navarea warning is superseded by a newer warning and is therefore cancelled
(2) A navarea warning self-cancels as it has a set cancellation date; and
(3) The navarea coordinator cancels the message.

The in-force list of warnings can be received through SAT C every week from which the OOW can determine the notices that are cancelled. Also, it is possible to establish which notices have been cancelled via Chartco by logging onto the Cancelled view. On receiving the in-force list of warnings, the OOW must first identify the cancelled warnings. The navarea file must then be updated, and the cancelled warnings segregated and put in the CANCELLED section. If paper charts are being used, the cancelled warnings that have been plotted on the charts must be erased along with the warning number that is written at the bottom of the chart. As we can see, navigational warnings play a major role in enhancing the safety of the ship. It is important to monitor navigational warning broadcasts prior to sailing. In some underdeveloped parts of the world, the coastal countries lack the facilities to broadcast navigation warnings for adjacent waters. A hazardous situation may exist and may go unreported unless a passing vessel's crew takes notice and makes a report to one of the WWNWS navarea coordinators. This not only improves the safety of navigation at sea but also helps prevents accidents from happening where they can be easily avoided.

Rogue waves

A ship at sea faces many different difficulties, especially in its motion. The forces acting on the ship is what maintains the equilibrium of the ship. Storms and heavy winds are known as factors which can affect a ship's equilibrium. However, there is one factor which is more dangerous than most and has the capacity to capsize a ship in no time at all – rogue waves. Rogue waves are one of the most terrifying issues faced by ships. Rogue waves, which are also called killer waves or freak waves, are dangerous for all types and kinds of ships. Rogue waves usually occur in deep water and have a focusing effect which is created by several waves joining together. The dangerous thing about these waves is that they appear without warning. Though rescue methods can be tried during the occurrence of these waves, they usually fail because at times these waves are so quick that precautionary measures aren't enough to counteract the effect of the wave. It should be noted that rogue waves are not caused by other phenomena such as tsunamis, earthquakes, or volcano explosions. Nobody really understands why or how rogue waves form, though expert consensus states rogue waves exceed 25 metres (82 ft) in height.

Figure 7.3 Winter, North Atlantic, 1958.

Rogue waves and the effects on ships

Bow slamming

When a ship encounters high waves (especially in head seas), high-amplitude pitching and heaving combined produces an effect that pushes the bow out of the water. As the wave passes aft ward, the bow falls onto the surface (slamming the surface), with high acceleration, resulting in tremendous slamming forces in the forward structure of the ship.

Propagation of cracks

Due to high slamming and pounding forces in the forward structure, the hull at the bow section is often prone to cracks that can propagate over the entire depth of the bow section.

Buckling of plates

The shell plates at the bow and the bottom plating up to 25% of the ship's length aft of the forward perpendicular are subject to effects of slamming which result in buckling of these plates. This is especially so with the bottom plating in the forward region. In most lading cases, the ship is in a hogging condition, which maintains the bottom shell in a state of compression. Major augmentation of stresses in the bottom plating, therefore, result in exceeding the buckling stress of the material, which may be much lower than the ultimate tensile stress.

Ultimate failure

When forward structures have been subjected to many cycles of freak waves or slamming forces over a longer period, the structure undergoes fatigue. If scantling and structural surveys are not carried out regularly, then ultimate failure, leading to a complete rupture of the bow sections, may happen when encountering freak waves. To avoid these problems from occurring, ship architects and designers have developed methods to combat freak waves by incorporating various factors of safety into the structural design. We will very briefly discuss these below.

Inclusion of rogue wave behaviour in structural formulae

When the scantlings of a ship are calculated in the preliminary design phase, designers use empirical formulae that are tested and recommended by the classification societies. These formulae have been developed through extensive observation and analysis of statistical data gained from studies of the stress ships are subjected to at sea. From these analyses, a factor of safety can be considered in determining the design and construction of scantlings, to prevent failure caused by waves that are above significant height. The hull girder is additionally strengthened at the bow. Some of the additional structures that are included are:

(1) Panting stringers that run longitudinally, are welded to the side shell forward of the collision bulkhead. The height between each subsequent stringer is usually 2 to 2.5 metres (6.5 to 8 ft)
(2) Panting beams that run transversely inside the deck shell, joining the panting stringer to the centre line wash bulkhead
(3) Angled pillars are used to support the panting beams at the centre line
(4) Panting web sections or perforated flats are used between every one or two panting stringers. The side frames are end-connected to these together with the panting stringers via tripping brackets to ensure smooth stress flow; and
(5) Plate floors are used at every frame space ahead of the collision bulkhead.

Navigational measures and bridge actions

Navigational measures must also be taken on board the ship to combat the effect of rogue waves, especially in high sea states. The purpose of adjusting navigational measures is just to reduce pitching amplitudes. The pitching amplitudes depend on the encountering angle and the encounter frequency. The encounter frequency is managed by changing the speed of the ship, and the encounter angle is altered by changing the heading of the ship. Though the latter may not always be feasible on fixed routes, the former is mostly used to reduce unwanted motions on the high side.

In this chapter we have looked at some of the main issues ship's navigators face as the OOW. In the next chapter, we turn our attention to the principles of Passage Planning.

8 Passage planning

Shipping cargo from one port to another involves the coordinated working of various operations both on land and at sea. One of the integral parts of shipping operations is passage planning, which is mainly undertaken by the ship's navigational officer. A passage plan is essentially a comprehensive, berth-to-berth guide, developed and used by the vessel's bridge team to determine the most efficient route, to identify potential problems or hazards along the route, and to adopt Bridge Management Practices to ensure the vessel's safe passage. SOLAS Chapter V, annexes 24 and 25, entitled Voyage Planning and Guidelines for Voyage Planning respectively, provide guidelines regarding the regulations pertaining passage planning. It is vital that all ship's navigators are fully cognisant of the demands SOLAS Chapter V places on them. In this chapter, we will discuss the main requirements and basic stages of passage planning.

Navigational charts

At the very centre of the passage planning process is the navigational chart. These charts are akin to maps of the sea floor. They are classified based on the regions where they are printed, published, and updated by regional hydrographic authorities. The leading world's hydrographic authority is the *United Kingdom Hydrographic Office* (UKHO),[1] which is led by the Hydrographer of the Navy, and operates under the command of the Admiralty. The Admiralty is the senior command of the Royal Navy. Of their various functions, one is to gather information in the form of hydrographic data, analyse it carefully, update it and disseminate the data in the form of paper and electronic charts and bevy of allied publications. A series of diagrammatic representations and indexes showing geographical limits of these charts are provided in the Catalogue of Admiralty Charts and Publications (NP 131). NP 131 divides the various maritime regions of the world into folios, which contain further detailed charts for each specific subregion.

UKHO and foreign charts

Though the UKHO has charted much of the world's oceanic regions and ports, several areas remain outside the scope of Admiralty Charts. For these regions, local hydrographic offices are responsible for preparing what are called Foreign Charts.

DOI: 10.1201/9781003291534-8

These charts are produced by local hydrographers using internationally recognised chart symbols and abbreviations. Whilst they are not produced or maintained by the UKHO, there is an expectation that they meet international standards and follow broadly the same process of production and update as UKHO charts. Another range of charts, which fall under the purview of Admiralty Charts, are Australian and New Zealand Charts. These charts are produced and updated by the *Australian Hydrographic Office*[2] and *Land Information New Zealand*[3] respectively. Though they are published under the name of the Admiralty, they carry Australian and New Zealand chart numbers instead of UKHO chart numbers. These charts are specific to the ocean areas surrounding Australia, New Zealand, the Islands of Macquarie, and all navigable waters up to and around Antarctica. Under a similar arrangement, Japanese charts are also published by the Admiralty but are drawn up and maintained by the *Japanese Hydrographic Association*.[4]

A detailed list of charts in the Japanese folio can be found in NP 131, and their corrections are provided in both Admiralty Notices to Mariners and as Japanese Notices To Mariners as well. There are two types of Japanese charts: one with the prefix W and the other with the prefix JP. The W prefix means the chart and associated notices are published for worldwide use in English and Japanese, and landmass colours are printed in grey. For JP charts, the language used is Japanese only with landmass printed in standard yellow. For Canadian and US waters, charts are produced according to the *Canadian Charts and Publications Regulations* and the *US Code of Federal Regulations*. Vessels navigating certain Canadian and US waters are required to only carry and maintain Canadian and US charts.

Although charts may be published by different hydrographic offices, they generally share a uniformity of symbols and abbreviations which are internationally recognised. This means irrespective of who produced the chart, they should be interchangeable and usable by navigators no matter what flag the ship sails under.

Chart scales

The scale of the chart is the first thing to check before starting the chart work. The chart scale is indicated below the name of the chart, for example:

THE SOLENT AND SOUTHAMPTON WATER
DEPTHS IN METRES
SCALE 1:25 000

After confirming the chart's scale, briefly glance along the latitude and longitude graduations that go up and along the sides of the chart. Once familiarised with the scale, the chances of error are markedly reduced. Also, it is good practice to take note of the chart's depth scale as well.

Small-scale charts

Small-scale charts show specific details such as light vessels, light floats, laybys, territorial waters, and land separating adjoining countries, obstructions, shoals, reefs, buoys, and fog signals. As the name suggests, they represent a vast stretch of an area on a small scale, thus showing macroscopic features. When preparing a passage plan, it is normal practice to use small-scale charts – especially for open seas such as transiting the Indian or Pacific Oceans – before transferring the passage plan onto a large-scale chart. This ensures that any navigational hazards are accounted for during the passage planning process.

Large-scale charts

Large-scale charts provide a much broader view of the sea area and tend to include microscopic details. An area covered by a small-scale chart can often be split into as many as 10, 20 or even 30 large-scale charts showing features such as principal lights, fog signals, light vessels, lighthouses, aids to navigation, leading lights, directional buoys, channel buoys, refineries, terminals, underwater cables, and pipelines, and so forth. The level of detail for each mark differs considerably between small- and large-scale charts. For instance, on a small-scale chart, a lighthouse is marked with a symbol only. On a large-scale chart, the lighthouse symbol is accompanied by the range, height, luminescence, and any other major characteristics of the lighthouse.

Chart notes

Once you are aware of the chart scales, read the notes which are also found below the chart name. These notes contain important information regarding navigation through the area covered by the chart. The information will include any submarine or naval exercise areas, anchoring, fishing areas, traffic separation schemes, and any additional information about wrecks, tidal information and local regulations concerning navigation.

Chart symbols

When looking at a chart for the first time, it is understandable that the vast array of symbols and marks littered across the chart will be confuddling. But understanding what these symbols and marks represent is fundamental to carrying out good-quality chart work. Most charts share a relatively small selection of symbols and marks, which, with study and experience, can be readily recognised. More obscure and unusual symbols and marks however can be identified by referring to NP 5011.[5]

Chart corrections

It is the responsibility of the Second Officer to keep all charts correct and up to date with the permanent and temporary corrections that are received weekly via Notices To Mariners. The permanent corrections are marked on the chart using magenta ink and temporary corrections are marked using pencil. It is very important not to miss these corrections when doing chart plotting. As far as practical, try to avoid plotting course lines near or above any corrections. Temporary corrections are more likely to be missed out as they are marked using pencil, which tends to fade with time. It is also good practise to check that the corrections are up to date. These are normally annotated at the bottom of the chart.   ·

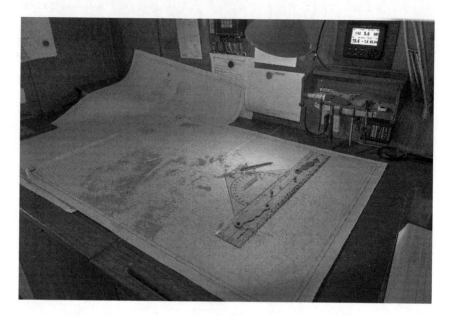

Figure 8.1 Example of an Admiralty Chart – Approaches to Trinidad.

Figure 8.2 Example of an Admiralty Chart – Cap Bon to Tòbruch.

Steps to passage planning

A ship's passage plan typically involves four main stages. These are:

(1) Appraisal
(2) Planning
(3) Execution; and
(4) Monitoring.

Each stage in the passage planning process has a purpose and it is extremely important to carry out each step correctly. At the start, a rough estimate is made of the entire voyage. Once the rough plan is ready, it is then modified and refined considering various details obtained from charts, the pilot book, weather routing, navigational warnings, and so on. These processes are carried out throughout the appraisal and planning stages. In the next two stages, i.e., execution and monitoring, the plan is used as a guideline. The voyage is executed as per the plan whilst taking into consideration any factors, such as weather, which may impact on or change the proposed route.

Appraise the passage plan

The first stage in passage planning is appraisal. At this stage, the master will discuss with the navigating officer (usually the Second Officer) how he intends to sail to the port of destination (though, in some cases, the master themselves may be required to plan the passage). This is the point of gathering the information relevant to the proposed passage. That information may be sought from a wide range of publications, including but not limited to any of the following: navigational charts, routing charts, NP 131 Admiralty Chart Catalogue, NP 136 Ocean Passages of the World, NP 45 Admiralty Sailing Directions, NP 87 Admiralty List of Lights and Fog Signals, NP 281 Admiralty List of Radio Signals, NP 201 Tide Tables, NP 120 Tidal Stream Atlas, NP 350 Admiralty Distance Tables, NP 100 Mariner's Handbook, D6083 Worldwide Load Line Chart, Chart 5011 Symbols and Abbreviations to be used on Charts, as well as Notices to Mariners and navigational warnings. Navigators should also call upon their own personal experience and knowledge of sea routes and prior passage plans. Once the navigator has taking into consideration the master's guidelines, the company's guidelines, ship's cargo, the general marine environment, and any other relevant factors that may affect the ship, the navigating officer draws a general track, which the ship will follow. For ease of planning, this track is first laid out using a small-scale chart, which is later transferred to individual larger scale charts.

Preparation

Having made a full appraisal of the proposed route, and using all information available, the navigator then begins to chart the vessel's passage plan. At this stage, the intended course of the ship is laid out on charts of suitable scale with all additional information marked accordingly. The plan is laid out from berth to berth, including any pilotage. It is a good practice to mark dangerous areas, such as nearby wrecks, shallow waters, reefs, small islands, emergency anchorage positions, and indeed any other information that may aid safe navigation. In addition to the above, it is strongly advisable to lay out the rate of turn for waypoints and the PI ranges for suitable objects, if any. Reporting areas should also be clearly marked on the charts. Further to marking out the track, the navigator should also account for any factors which may aid or hinder the vessel during its voyage. An example of the sorts of things that ought to be included are no go areas, margins of safety, charted tracks, course alterations and wheel over points, parallel indexing, aborts, and contingencies, clearing lines and bearings, leading lines, tides and currents, anticipated changes in engine status, minimum ukc, usage of the echo sounder, head marks, and natural transits.

It is worth mentioning at this point the importance of aborts and contingencies. When approaching constrained waters, the vessel might be in a position beyond which there is no possible action but to proceed. For instance, where a vessel enters an area so narrow that there is no room to return. It is for this purpose that a position is drawn on the chart showing the last point where the passage can be safely aborted. With contingencies, the bridge team must always be aware that events might not go as planned and that emergency action may be required. Contingency plans account for such eventualities and are clearly shown on the chart so that the OOW can take swift action if required. Contingency planning includes marking alternative routes, identifying safe anchorages, waiting areas, and emergency berths.

Execution

Once the passage plan has been appraised and prepared, it is given to the master to sign off. The master will evaluate the passage plan and either approve it as it is or make changes based on their own experience and appraisal. Once the master approves the passage plan, it is time to execute it. At the execution stage, the navigating officers carry out the plan as it has been prepared. Following departure, the speed is adjusted based on the ETA and the expected weather and oceanographic conditions. The speed should be adjusted such that the ship is neither too early nor late at its port of destination. The master must also account for the volume of potable water, fuel, and food onboard.

Monitoring

Whereas the execution phase involves putting the plan into action, the monitoring phase involves an ongoing and continuous reassessment of the voyage. In an ideal world, the vessel would be able to follow the passage plan from one berth to the next without any changes. Unfortunately, the world we live in is far from ideal; therefore, even the best-prepared and thought-out passage plans are likely to change at some point. There are various ways that navigators can monitor the ship's progress underway. Parallel indexing, for example, can be used to maintain safe distances alongside hazards to navigation.

Dead reckoning

Before the advent of modern navigational aids such as GPS and ECDIS, positions at sea were mostly determined through celestial and terrestrial methods, especially with the use of fixes from radio navigational stations. Thanks in part to the 32- point magnetic compass, ocean navigation has become much easier and accurate. Dead Reckoning (or DR) is one of the oldest and most trusted methods of navigation. In fact, dead reckoning remains the basic science underpinning modern-day navigation. Over two hundred years before the Italian explorer Christopher Columbus made way to the Americas in 1492,[6] Mediterranean navigators developed a technique called dead reckoning. In fact, the Carta Pisana, which dates back as far as 1275, is the oldest-known dead reckoning chart. Dead reckoning (or DR) as it is usually referred, is the process by which the ship's current position is calculated based on a previously obtained position. A previously determined position is also referred to as a fix; the estimated speed and course are applied to that fix to obtain the DR. To make things slightly simpler, a fix is defined as a position derived from an external reference point such as a position line. Two or more intersections among the position line provides a relatively accurate position of the vessel. The presence of a 'cocked hat' on obtaining and intersecting three position lines indicates an error. The optimum angle of intersection between two position lines is 90 degrees while that among three position lines is 120 degrees. The intersection is necessary to obtain, as a single position line will only provide an estimated position, i.e., the vessel is somewhere along that line. It is imperative to understand the concept of the fix and position lines to understand the process of obtaining the DR.

Whilst calculating DR is no longer needed as dead reckoning has been replaced by electronic aids to navigation and ECDIS, it is always good to develop and maintain manual navigation skills in case the electronic systems onboard fail. To simplify the concept of DR further, it is the process of calculating the current position by applying the speed and course to a previously calculated position. This method of calculating the vessel's position also enables us to calculate an estimated future position based on the ship's current course and speed. Although DR is relatively accurate, there is an order of precedence when it comes to manual plotting. First is the fix, followed by the running fix, then the estimated position, and finally dead reckoning. When plotting a DR on the chart, it is important to keep in mind that every time there is a change in course and/or speed, the DR position must be replotted. If the predicted course line is judged or found to be incorrect, then again, the DR must be replotted. When doing so, remember not to draw a new course line from an Estimated Position (EP) and always plot a new course line from a fix. There are bound to be errors in any data obtained based on estimation, as with DR. In determining a future position, or even an assumed current position, the factors of set and drift of current, wind and other external factors are not accounted for. This means that whilst the estimated plot will be within a reasonable margin of error, it will not be entirely accurate.

Before the invention of modern navigational technology and accurate hydrographic charts, sailors measured the distance travelled by their ship from a fixed point such as their home port or harbour. Each day they would be keep a log.

A chart was prepared using pins to mark points on it. As the ship passed each point, this would be recorded in the log. The ship's speed was measured using a chip log. This was essentially a piece of wooden log which was thrown into the sea ahead of the ship's bow. The forward and aft of the ship was marked with some form of indication. As the ship sailed past the log, a note was made of when each indicator was passed. The speed of the ship was then calculated using the time it took for the log to pass the two marks. These measurements of course and speed were recorded on the chart and in the log and were updated at regular intervals. Whilst these techniques are, by comparison to today's standards, very rudimentary, they were effective enough to last well until the mid-17th century when navigation techniques and equipment started to become more technologically advanced. Even today, survivalists recommend having an appreciation of such techniques in case the worst happens and the crew are forced to abandon ship.

Position plotting is a vital element of passage planning and continuous monitoring. Safe navigation is contingent on accurately positioning the vessel. When sailing across open seas or through coastal waters, narrow canals, rivers or even into ports and harbours, the key requirement is to keep accurate position fixes. Various methods are available for position fixing, including visual bearings, radar fixes, terrestrial and celestial observations, and GPS fixes. Each of these methods can be used alone or combination to verify each other. For example, in open seas where traffic density is scarce, a celestial observation or GPS fix is an acceptable method for position fixing; however, in coastal or congested waters or in traffic separation schemes, narrow channels, canals, and river passages, radar, visual or terrestrial bearings are more reliable compared to GPS fixes.

Parallel indexing

Parallel indexing is a technique where a line is drawn on the chart or radar screen parallel to the ship's courses but offset by some distance to the port or starboard side. This parallel line allows the navigator to maintain their course but at a safe distance from any hazards. The basic principle of parallel indexing is to maintain and follow a particular course – i.e., a bearing line – which is drawn parallel to the original course with a known and fixed perpendicular distance between both lines as a reference. The increase or decrease of the perpendicular distance between the bearing lines drawn parallel to the course line and the ship's position at any time will indicate cross-track deviation from the initial planned course and thus advise whether the ship is falling out of a traffic lane, entering a traffic separation zone, or closing in on a navigational hazard. The reference point from which the bearing line parallel to the course line is drawn is taken as a fixed buoy, lighthouse, headland, jetty, fixed platform, or some other fixed and conspicuous object. Therefore, the imaginary line drawn parallel to the course line is always at a fixed distance. While the ship follows the course to steer, parallel indexing ensures the vessel always remains at a fixed distance from the hazard. In summary, parallel indexing is a navigational method that alerts the navigator that they have steered to close to a hazard. For further information and guidance related to parallel indexing, refer to SOLAS Chapter V regulation 34, IMO resolution A893[7] and the OCIMF Guidelines.[8]

GPS fixes

GPS positions can be plotted directly on to all new Admiralty charts as they remain in the WGS-84 Datum. GPS positions have an accuracy of around 50 metres (165 ft). Always try to avoid plotting GPS positions when navigating in coastal and restricted waters as such margins can be inherently dangerous. Also, when navigating in shallow waters surrounded by high cliffs and mountains, multi-path errors can occur, resulting in reduced accuracy.

Radar fixes

Radar fixes are considered one of the most reliable types of fixes when navigating in coastal waters. Radar fixes are most effective when fixed using radar-conspicuous objects such as a racon or ramark. Racons and ramarks are indicated on charts by a magenta circle and are easily identifiable on the radar itself. It is strongly recommended to take note of racons and ramarks when passage planning and to use them for position plotting. When plotting course lines, marking out radar-conspicuous objects which can later be used for fixes will help save time and make position plotting much easier.

Visual fixes

Visual fixes give reliable positions when navigating close to shore. They can be taken from lighthouses and other static aids to navigation. The bearings from these objects can be obtained using an azimuth ring. Such bearings should be quickly and carefully plotted on the chart for best accuracy. If three bearings are used, the position will be inside the resulting 'cocked hat'. In best practise, however, bearings from objects at right angles to each other provide the best visual fixes.

Position circle and position line

The position circle is a circle drawn from an object at its centre, with its range calculated as the radius to the position of the ship. The position line is a line drawn from the object along its bearing from the ship. The ship's position is known is to be along that line. A fix can be obtained using any two parameters, i.e., two position lines or two position circles/one position line and one position circle. The more parameters that are used, the better the accuracy of the fix. Since the ranges obtained from radar are the most accurate, three position circles can give the best fixes. When there is only one object available, it is possible to fix the vessel's position using the running fix technique. Whatever the case, when doing chart plotting, the navigator should be familiar with different fixing techniques and use them as appropriate for the conditions and available resources. Chart work is a skill that requires accuracy and precision. Safety of navigation depends on the quality and reliability of chart plotting. Hence, navigators doing chart plotting should do it with care and attention. A wrong course line or position can mislead the vessel, possibly leading to accidents. Just as an over-reliance on GPS is a mark of poor seamanship, so is a lack of understanding and ability in different chart work techniques.

Chart corrections

Whenever a chart is amended, for example to account for changes in charted depths or the repositioning of buoys, a revised chart will be issued by the Admiralty. On the other hand, if wholesale changes are made to an existing chart, such as the addition of a new channel or the imposition of traffic separation schemes, then a new edition will be published which replaces all extant charts under that number. NP 131 Catalogue of Admiralty Charts and Publications is published every year with the current and updated editions of all charts. Any amendments that occur during the year are published and circulated via the Weekly Notices to Mariners. These will indicate any new editions or new charts that must be inserted into the chart catalogue by the navigating officer. The chart catalogue consists of a folio system comprising the Chart Correction Log and Folio Index as well as the NP 131 Admiralty Chart Catalogue, NP 45 Admiralty Sailing Directions, NP 87 Admiralty List of Lights and Fog Signals, NP 281 Admiralty List of Radio Signals, NP 201 Tide Tables, NP 350 Admiralty Distance Tables, and Chart 5011 Symbols and Abbreviations to be used on Charts.

When compiling and updating the chart catalogue, the following points should be considered. First, the chart correction log should have details of all charts present on board with an index of corrections (temporary and preliminary) mentioned against each chart number. All charts and publications are to be corrected on a weekly basis in accordance with the instructions provided in the Weekly Notices To Mariners. The correction numbers should be entered in the same index mentioned above against the chart numbers respectively, so that it can be used as a quick reference to check the status of any corrections for each chart. Secondly, there should be a designated chart supplier or service provided who can supply new and updated charts and publications in a timely manner. Third, all ships must carry and use equipment for receiving navigational warnings and NAVTEX communiqués. Fourth, whenever a new chart or new edition is published, a Weekly Notices To Mariners is issued. On receiving a new edition or new chart, the chart correction log must be updated with the relevant folio number against the chart number in the index. Either 'NC' or 'NE'[9] should be annotated adjacent to the chart number in the correction index. Once the chart has been logged, it should be checked for any pending temporary and preliminary corrections and or navigational warnings.

On receiving a chart or notification through Weekly Notices To Mariners replacing a previous extant chart, annotate the chart correction log with 'replaced by' or 'cancelled by' against the mentioned chart. The relevant folio number should then be inserted on the replacement chart once received. NP 131 Admiralty Chart Catalogue, Weekly Notices To Mariners, Cumulative Notices to Mariners, and the

Annual Summary of Admiralty Notices to Mariners are important sources of information for keeping the chart folio system updated and correct. For guidance on correcting charts, the Admiralty has published a guide – How to Correct Charts the Admiralty Way[10] – which describes the various forms of corrections applied to charts and best practices for maintaining chart folios. The importance of keeping charts current and up to date cannot be stressed enough as the following real-life incident demonstrates due to the negligence of the navigating officer, a vessel was refused entry to the Port of Baltimore in the United States. After two days, special permission was granted by the US Coastguard (USCG) upon request by the ship's agents. The vessel did not have the latest edition of approach charts for the Port of Baltimore. The new edition had been published four weeks prior. The vessel had received notification of this but had made no effort to procure the latest editions. The USCG, during their annual inspection of the vessel, discovered the error and fined the vessel for non-compliance. It took a further two days for the ship's agent to provide the vessel with the correct approach charts, after which the vessel was permitted to berth. It is therefore important that when updating and correcting the ship's charts, due diligence and extreme care is taken. In another example, near the Batu Berhanti Light in the Singapore Straits a wreck is marked and surrounded by four cardinal buoys. During chart corrections the notice read 'Insert Isolated Danger Mark' adjacent to the light. The navigating officer misinterpreted the notice as 'Delete'. Instead of inserting the notice, the navigating officer deleted the mark. The ship's course was laid from that area and while transiting it was observed that the vessel passed very close to the Isolated Danger Marks which were wrongly indicated on the chart. Just a few metres either side of the marks and the vessel would have suffered a serious accident.

Chart correction software and computerised charts

As the maritime industry moves forwards, Weekly Notices To Mariners are available online using the software platform Chartco. Chartco performs the same function as the physical chart folio; instead of distributing printed hard copies, however, Chartco provides a web-based database and chart corrections service. Corrections are listed according to specific chart folio numbers. This means the navigating officer can access the system by typing in a specific chart number to print and apply any specific or all applicable corrections for any chart present onboard.

Notices to Mariners

There are five primary sources of Notices To Mariners:

(1) Weekly Notices to Mariners
(2) Cumulative Notices to Mariners
(3) Annual Summary of Notices to Mariners
(4) Navarea warnings; and
(5) NAVTEX warnings.

Weekly Notices to Mariners

Issued on a weekly basis by the Admiralty, these Weekly Notices to Mariners contain corrections to charts and various other publications such as the NP 40 Admiralty Sailing Directions and NP 285 Admiralty List of Lights and Radio Signals. They also include Temporary and Preliminary Notices to Mariners for correcting charts. The year begins with notice No. 1 being issued in the first week of January. The numbering increases chronologically up to the last week of the year. They also notify ships of any new charts or new editions being published and the cancellation and replacement of existing charts. The corrections are mentioned either in the form of coordinates with instructions or in the form of overlay chart tracings and blocks to be cut out and pasted. New editions of NP 74 Admiralty List of Lights and Fog Signals, NP 40 Admiralty Sailing Directions, and NP 281 Admiralty List of Radio Stations are promulgated as and when they become available.

Cumulative Notices to Mariners

These are published twice a year in the months of January and July. The main purpose of the Cumulative Notices to Mariners is to provide a list of current editions of all extant Admiralty Charts and the serial numbers of notices affecting them which have been issued in the preceding two years. The charts onboard can then be verified against the cumulative list for any missing corrections.

Annual Summary of Notices to Mariners

All Temporary and Preliminary Notices to Mariners for a previous year are published as an abridged version in the Annual Summary of Notices to Mariners. This document provides a quick reference for navigators to compare and confirm all temporary and permanent corrections applicable to a specific chart. It also contains a summary of corrections to NP 40 Admiralty Sailing Directions.

Navarea warnings

Navarea warnings are promulgated to ships operating in specific areas. As we mentioned previously, the world is divided into 21 navigational areas (or navareas) and these warnings are broadcast by the Worldwide Navigational Warning Service. Through a network of Area and National Coordinators, important international, coastal, and local warnings and notices related to safe navigation are transmitted directly to ships which are received on MF and HF radio frequencies, and satellite networks.

NAVTEX warnings

These are broadcast by Telex Broadcasting Stations located in various coastal areas. Mostly, they broadcast navigational, meteorological and distress warnings. They provide a vital source of early warning to ships engaged in coastal navigation.

In this chapter, we have covered some of the core concepts and principles of passage planning. Passage planning is – in itself – a detailed and complex process which requires skill and practice. Even so, it is a core duty of the ship's navigating officer and the OOW to be able to develop a passage plan and plot fixes as the ship makes way to its destination. In the next chapter, we will look at some of the other duties of the OOW.

Notes

1 https://www.gov.uk/government/organisations/uk-hydrographic-office
2 https://www.hydro.gov.au/
3 https://www.linz.govt.nz/data/linz-data/hydrographic-data
4 https://www.jha.or.jp/en/jha/purchase/ntm.html
5 NP 5011 Symbols and Abbreviations used on Admiralty Paper Charts
6 Historians continue to argue whether Columbus or not discovered America, though it is widely accepted he never set foot in North America. During four separate trips starting in 1492, Columbus landed on various Caribbean islands that are now the Bahamas as well as the island later called Hispaniola. It is also accepted that he explored the Central and South American coasts.
7 https://wwwcdn.imo.org/localresources/en/KnowledgeCentre/IndexofIMO Resolutions/AssemblyDocuments/A.893(21).pdf
8 Guidelines for Offshore Tanker Operations, 1st Edition, 2018.
9 NC – New Chart; NE – New Edition.
10 https://www.admiralty.co.uk/AdmiraltyDownloadMedia/ADMIRALTY%20 Leisure%20Charts/Guidance/Keeping%20ADMIRALTY%20Leisure%20Charts%20 up-to-date.pdf

9   Officer of the Watch

The navigational watch is one of the most significant shipboard operations per-formed by deck officers. When the ship is at sea, the bridge is the only location that is perpetually manned. This is not a moment in time where the bridge can be left unattended, even if just for a few minutes. As the navigating officers must keep bridge watches round the clock, the practice of taking over the watch by a relieving OOW from the outgoing OOW is performed twice daily. When taking over the watch, the oncoming OOW is acquainted with the ship's position, speed, and course; traffic density; past, present, and expected weather and sea conditions; the state and condition of bridge equipment; logbooks, checklists, and daily orders; readiness of the lookout and or helmsman; any activities taking place on deck or in the engine room; and any conditions that may require the master to attend the bridge. Whilst on watch, the OOW is responsible for complying with the COLREGS and with all standing orders issued by the master. In summary, the OOW has three primary duties: perform safe navigation; maintain watchkeeping; and monitoring the GMDSS.

In this chapter, we will briefly discuss each of the main duties of the OOW and examine what these duties entail.

Taking over the watch

By far, the most important action the incoming OOW can do is check the position and speed of the ship. Once satisfied with the ship's position, as per the ship's chart, it is a good practice to browse through the preceding chart(s) and the succeeding chart(s) to become acquainted with the course to be plotted over the following watch. Similarly, check for any waypoints, course alterations, any reporting points, traffic separation schemes, shallow patches, or any dangers to navigation along the intended track marked on the chart and compare the course on the chart with the course as it appears in the passage plan.

Checking traffic density

Once satisfied with how the vessel is progressing, it is necessary to check the traf-fic density immediately around the vessel. This can be done by first checking the radar, and then by scanning the horizon from both bridge wings. Remember to

DOI: 10.1201/9781003291534-9

check the stern as well. If at the watch handover the ship is already engaged in overtaking another vessel, is in close quarters, or is in a crossing situation, do not take over the watch until the situation is resolved, and the other vessel(s) are past and clear.

Weather conditions and night vision

The next important factor to check when taking over the watch is the current weather condition. Always note wind speed and direction, and the set and drift of the current. If possible, try and determine the likelihood of precipitation or restricted visibility occurring during the watch. During the hours of darkness and or restricted visibility, it is usually necessary to adjust the lighting on the bridge to a low setting. This makes it easier for the eyes to adjust to the darkness outside. Bear in mind that it takes on average 15 minutes for human eyes to adjust from light to darkness.

Bridge equipment and dimmers

When taking over the watch, ensure all bridge equipment is ready and operational. Any troubleshooting or problems with the equipment should be passed on by the outgoing OOW. During the preceding watch, it is likely the outgoing OOW set the bridge equipment and radar to suit their own preferences. Always check these at the start of the watch and make the necessary adjustments, if needed. When sailing at night, the bridge display equipment should be set to night mode and all other equipment dimmed to prevent back scatter.

Logbooks, checklists, and daily orders

At the start of every watch, the incoming OOW must check and initial the ship's log, the relevant checklists, daily orders, night orders and the master's standing orders. Any concerns with the previous watch should be raised with the outgoing OOW before signing the logbook and before the outgoing OOW has left the bridge. By signing the logbook, you are confirming that you are now in command of and are responsible for the ship, its crew, and its cargo.

Readiness of the lookout and helmsman

It is occasionally a requirement to assign an able seaman as part of the bridge team for look-out duties. The lookout may also be assigned the responsibility of manning the helm for manual steering, such as emergencies, landfall, and when approaching port or transiting through restricted waters. At all other times, it is not uncommon for the OOW to be the only person on the bridge performing lookout duties. In these situations, the lookout/helmsman should be put on notice and provided with a portable radio in the event their presence is needed on the bridge at short notice.

Miscellaneous activities on deck or in the engine room

When taking over the watch, it is vitally important for the outgoing OOW to inform the incoming OOW of any activities taking place on deck or in the engine room. These activities may impact on the ship's ability to perform under emergency conditions; therefore, the OOW must be made aware of what is always happening in parts of the ship. Miscellaneous activities are those which do not fall within the day-to-day duties of the officers and crew.

These might include:

(1) Tank entry, inspection, and cleaning
(2) Cargo hold entry
(3) Bilge well entry and/or routine alarm checks; fire watch, hot works, or welding on deck
(4) Working aloft
(5) Working on the monkey island
(6) Working on the masts; working overboard; and
(7) Drills.

The relieving officer has the authority to demand any information regarding the navigation and progress of the vessel up to and including the point where the outgoing OOW signs the logbook, which completes the watch handover.

Duties of the OOW after taking over the watch

Once the outgoing OOW has signed the logbook, they are no longer responsible for the safe navigation of the vessel during that watch, notwithstanding their duties and responsibilities as a deck officer. Whilst keeping a watch on the bridge the officer is the representative of the ship's master and has total responsibility for the

Figure 9.1 OOW on the bridge.

safe and smooth navigation of the vessel. They are responsible for ensuring the ship complies with SOLAS, the COLREGS, and regulations relating to the navigation of a ship at sea. These responsibilities are broadly defined into three core areas:

(1) Navigation
(2) Watchkeeping; and
(3) GMDSS.

We have already covered the principles of passage planning and plot fixing so we do not need to concern ourselves with these duties here. Rather, as well as maintaining the ship on course, the OOW has several primary duties which can be summarised thus:

(1) *Compare the compasses.* This is done to determine the 'precise estimated window' within which compass errors can affect the course of the ship. In the event the gyro compass fails, the OOW must be aware of the extent to which the error of the magnetic compass might affect the course being followed. Moreover, a comparison of the repeaters is essential to ascertain whether the repeaters are aligned with the master gyro and showing the correct reading which is needed when calculating the compass error using the azimuth.

(2) *Check soundings by the echo sounder.* The ukc and depth of water at any point is critical for safe navigation. Whereas the master may request, at their discretion, a record of the present depth, it is important for the OOW to know the current water depth to avoid running aground or stranding

(3) *Ensure the lookout is alert.* Rule 5 of the COLREGS places special emphasis on the need for lookouts and to that effect states "every vessel shall, at all times, maintain a proper lookout by sight and hearing, as well as by all available means appropriate in the prevailing circumstances and conditions so as to make a full appraisal of the situation and of the risk of collision". This must not be misinterpreted with rule 19 of the COLREGS which mandates additional lockouts during periods of restricted visibility. Rule 5 quite clearly mandates that a lookout must be always present on the bridge, irrespective of the prevailing weather or sea conditions. As mentioned previously, the role of lookout may be delegated to the OOW themselves unless an additional lookout is necessary.

(4) *Checking the ship's position.* When taking over the watch, the incoming OOW must check the position plotted by the outgoing OOW and to make themselves satisfied that the ship's position is accurate

(5) *Draught.* The ship's draught must be displayed on bridge and updated whenever there is a change. Remember, the ship's draught has a direct relationship with the ukc, and an erroneous draught may lead to the vessel grounding or stranding

(6) *Check the gyro and its error.* Most, if not all, of the equipment on the bridge will have some element of error. Whilst it is important that they are all factored in, the gyro is one piece of bridge equipment that must be always checked and confirmed in good working order

(7) *GMDSS.* The GMDSS watch is crucial to the crew's safety and must be maintained on the stipulated frequencies as per regulations. Additionally, all MSI promulgated via NAVTEX, SATCOM, EGC or VHF must be checked and logged irrespective of whether such information directly' or indirectly impacts on the ship

(8) *General rounds of the ship.* Soon after handing over the watch, the relieved OOW may carry out rounds of the ship to confirm that fire safety has been maintained, there are no signs of breach or indeed anything else out of the ordinary. If rounds are completed, the outgoing OOW must inform the current OOW that such inspections have been performed and that nothing is amiss.

Different entries to be made in the ship's logbook

Being a deck officer involves myriad responsibilities that are spread across the different aspects of the vessel's operation. It is a challenging job, yet a rewarding occupation for those willing and able to put in the commitment. For this reason, every task for every department must be recorded in a logbook. The Deck Logbook is the record of action, event, incident, or occurrence that takes place on the ship's bridge. The logbook is used to record various data, scenarios, and situations (including emergencies and actions) which may later be used for reference, case study and for marine insurance purposes or legal investigations. Although supplementary logs are kept in addition to the Deck Logbook, it is the latter which carries the most weight and must be kept current and accurate. During an average watch, the OOW may complete the logbook as many as a dozen times with all sorts of ephemeral information. There are, however, entries which must be recorded during every watch, including:

(1)   The position of the ship's latitude and longitude at different intervals
(2)   Times when navigation marks are passed
(3)   Times, details, and reasons for any course alterations
(4)   Details of meteorological and weather conditions, including sea conditions and swell together with wind speed and wind direction readings
(5)   Details of the ship's movements, including rolling, pitching, and heaving
(6)   Details of any abnormal conditions
(7)   The ship's speed in knots and rpm
(8)   Details of any kind of accidents such as stranding and grounding
(9)   Details of any physical contact with floating objects
(10) Details of distress signals received or issued
(11) Details of what assistance was given or received in relation to distress signals
(12) If salvage operations are performed, a complete record must be kept
(13) If there is an oil spill or other pollution-related incident from the vessel, complete details must be recorded including the ship's condition, position and the time of the incident
(14) A record of the general watch routines performed, including fire watches
(15) Estimated and actual times of arrival and departure to and from port
(16) If anchoring, the actual time of lowering and heaving the anchor

(17) Any heading and compass errors
(18) Any drills and training carried out as well as stowaway and security-related inspections
(19) Records of stores received; and
(20) Any other entries as required by the master, the company, and Flag state administration.

When completing the logbook, original pages must never be removed. This is because the logbook serves as an official record of evidence should the vessel be involved in an accident or incident. Moreover, only official designations and symbols may be used and if there is insufficient space in the Remarks section for a full account, secure an additional piece of paper instead of writing over text or into the margins. Remember, each logbook entry carries the OOW signature, which means the OOW is responsible for all that is recorded by them. Any erroneous or factually incorrect information that is knowingly made by the OOW is a serious offence punishable by fines and even imprisonment.

Standing and night orders

Standing orders are a set of instructions provided by the master to ensure safe ship navigation and operations whether at sea or in port. These instructions encompass a wide area from navigation to the rules of conduct of officers. Standing orders must always be followed and are signed by every officer on board, making them liable. Night orders are a supplement to the Standing orders and come into force as the master proceeds to take rest during the night. The Standing orders remain in force and the night orders add specific points be followed when the master is unofficially off duty. The master writes the night orders every night, with specific regard to the prevailing state of the weather, sea conditions, and marine traffic. These are generally handwritten and signed by each OOW. The night orders are an important document as the master will use their extensive experience and expertise to determine safe navigation during the hours of darkness. Although the master is unofficially off duty, they remain officially in command of the ship and are responsible for the safe passage and conduct of the vessel, even in their absence from the bridge.

Common mistakes ship navigators make that can lead to accidents

There is an old anecdote about a ship going up-river to a port when another ship is sighted on the radar. There is already a marine pilot onboard who informs that there is no mention of outbound traffic and the other ship on the radar is probably at anchor. To confirm what the pilot has said, the OOW calls the other ship on VHF and asks:"vessel in position ABC, are you a tanker?" to which the other vessel replies,"yes, I am a tanker."[1] Whilst this might seem mildly amusing, it is potentially

---

1 "What the OOW meant to say is "are you at anchor?" The other vessel interpreted this question as "are you a tanker?" This highlights the importance of clear and unambiguous communications, especially over radio and between vessels and shore-based operators.

the beginning of a very dangerous situation. The maritime industry is truly global, and the people who work on both shore and at sea come from a vast variety of nationalities and backgrounds. Some have a better grasp of English – as the above anecdote shows – which is the recognised language of the sea. The use of VHF for collision avoidance has always been debatable and whether or not to use it will depend on an assessment of each specific situation. The use of VHF is not the only problem the OOW faces during the long and tedious hours spent alone on the bridge. Whilst it would be impossible to list every mistake or eventuality that could lead to accidents, it is worth noting some of the most common so that these can hopefully be avoided.

First, never rely solely on radar. In one real-life incident, the master of a container vessel saw a small fishing boat near the bow of his ship. He rushed up to the bridge, immediately engaged the hand steering and in so doing avoided a collision by a matter of minutes. In this case, the OOW had not sighted the fishing boat as he was navigating solely by radar and this target was not picked up due to its size. At the time the incident occurred, the OOW was altering course for another ship that had been picked up on radar! Had the OOW kept a proper visual lookout by looking out of the bridge windows as well as using the radar, the situation could have been avoided entirely. There is a mantra that experienced seafarers like to remind their younger colleagues: nothing beats the "Mark 1 Eyeball method of navigation". Despite the rapid rate of technological advancement in virtual aids to navigation, this still holds true. A trained human eye can pick up much more information and the human brain can process it faster than any complex algorithm or radar.

Second, radar targets must always be verified visually. In an area of restricted visibility and dense traffic, the lookout on duty informed the OOW about a target that was sighted right ahead on the radar. The OOW did not take any action as he assumed that the target was either a fishing boat that would move away as the ship closed in, or it was a false echo. This resulted in a collision with another ship that was stopped and drifting. In this case, altering course even for a "false echo" would have been the best action to take. Safe navigation using radar can be done only when the navigator is confident in his understanding of the equipment and appreciates its limitations. Before radars were widely used on oceangoing vessels, there was a term called "Fairweather practice", and it is still relevant today. This meant that best use was made of the opportunity to use radar whenever it was available, combined with a mental image which would help in deciphering the radar picture during restricted visibility. Now with increased AIS dependence, there is even less use made of the ARPA function of radar. It cannot be stressed enough that the radar target must always be verified visually.

Third, never solely depend on the SMS. When approaching port having plotted the ship's position on the chart, the navigating officer informed the master that the vessel was north of the intended track, and that to correct it, the vessel should come south. Based on this information, the master ordered the ship to adjust its course accordingly. In so doing, however, the radar picture of landmarks and the navigator's assessment did not match. So, the master checked for himself and realised that the OOW had plotted the latitude incorrectly. Incidents like this are easy to do when the navigator is inexperienced or incorrectly interprets the SMS. When in doubt, apply the following rule of thumb: between two successive position plots, the vessel should not be in danger. Whenever a navigator plots position on a chart, it should be followed with a DR position, which will give a rough estimate of the vessel's location at the time of the next plot. Based on DR, the frequency of plotting can be increased or decreased.

Fourth, identify buoys correctly. An experienced chief officer was in the process of picking up the pilot in a buoyed channel. He was a little nervous as he was being assessed for his ship handling skills by the master. After entering the channel, the ship started setting and a starboard side lateral buoy was sighted right ahead. Seeing the buoy right ahead, the chief officer immediately ordered hard to starboard; the master overrode his order, realigned the ship again and handed the conn back. Somewhat like the east west north south cases mentioned above, this can happen during a lapse of concentration resulting in confusion regarding the colour of buoys and on which side to pass them. A common reason for this to happen is due to the different colours of buoys in IALA regions (see Chapter 2). Confusion such as this can be minimised by remembering can to port and cone to starboard when inbound, and reverse when outbound.

Fifth, always monitor the rudder angle. In a busy traffic separation scheme, the master had the conn of the vessel and the OOW was communicating with VTS when the lookout reported a stationary fishing boat on the starboard side had started moving and was crossing the bow. The master verified the movement of the target on the radar and ordered to helmsman "starboard 20". At this time, the bridge telephone rang and was answered by the master. After completing the telephone conversation, he looked up and noticed that the fishing boat was still not clear of the ship's bow. The master ordered "hard at starboard". At this time the bridge team realised that although the helmsman was repeating the order, he was applying the helm in the opposite direction. As this incident shows, it is vital to monitor the rudder angle indicator (RAI).

Operating the main engines from the bridge

As stated in the COLREGS regarding the use of ship's engines, and the SOLAS guidelines for Bridge Watchkeeping, duty officers must be fully aware of and conversant with the various bridge control systems and the use of the main engine when manoeuvring. To avoid making the mistakes discussed above, the OOW should always:

(1) Check the communication systems. First and foremost, the important thing before using the main engine is communication between the bridge and the engine control room or the local manoeuvring platform. The fixed communication system, hand-powered telephones, talk back system – and portable radios if used – should always be checked they are in good working order.

(2) Clear the propeller. Before trying out or operating the main engines it must be ensured that the propeller is clear of any nets, fishing lines, ropes, trawls, or any other unwanted objects. Often while the vessel is stopped, lying at anchor, berthed alongside, or moored at single point mooring, there are tug ropes in

the water, fishing nets or broken lines floating, which may get entangled in the propeller fins causing damage. Subsequently, any hazards such as these must be cleared before rotating the propeller.

(3) Ensure the engines are tried from the engine control room before taking over. It is very important to ensure that the propulsion plant is in good condition so that no issues are experienced when the engine is running during manoeuvring. This is best done by the ship's engineers who can assess the engine by trying for ahead and astern directly from the engine control room before transferring control of the engines to the bridge.

(4) Test the telegraph transmitter. Before operating the main engine, the telegraph transmitter from the bridge must be tested with the engine control room, and its operation confirmed at both locations for starting, stopping, reversing and speed setting. All indicators, push button switches, lamps and buzzers on the main engine telegraph panel should be tried out before starting the main engine.

(5) Transfer the controls and check the indicator lamp. The controls for operating the main engine must be provided or transferred through the command switch on the bridge telegraph panel. When changing over the control position to the bridge from the control station, operate the telegraph transmitter so that the indicator lamp of the handle match position lights up. The navigating officer must be aware of the emergency starting interlock if the emergency shutdown is in a non-operating condition.

(6) Put the telegraph transmitter into operational position during harbour passage. During harbour passage, the speed control of the main engine on the telegraph transmitter can be put into an optional position. The main engine speed, corresponding with the telegraph position, is set due to governor control. The position of the telegraph transmitter is set into the centre of each instruction slot; however, engine speed can be changed by adjusting it into any other position.

(7) Control the acceleration of the main engine gradually when the telegraph is put into NAV FULL. When the telegraph transmitter is put into the NAV FULL position, which is at the starting position of a load programme, the acceleration of the main engine is then controlled gradually and not instantly as per the position of the telegraph transmitter. The indicator lamp – LOAD PROGRAMME – then lights up during this command. For reducing the speed of the main engine, the telegraph transmitter is adjusted from the starting position of the load programme which reduces the speed from NAV FULL.

(8) Use PROGRAMME BYPASS for an instant reduction or acceleration in speed. In the event where a situation demands an instant reduction or acceleration of the main engines from NAV FULL, a PROGRAMME BYPASS switch is provided. When this switch is operated, it by-passes the LOAD PROGRAMME by reducing or increasing the engine's revolutions and thus brings the main engine speed instantly to the required position set by the telegraph transmitter. By pushing the switch again, the PROGRAMME BYPASS switches back to normal operation and the engine condition is restored.

(9) Know the manual emergency shutdown procedure. Manual emergency shutdown of a main engine is provided generally at following locations, e.g., the bridge and the engine control room. Manual emergency shutdown can be operated from any of the positions regardless of the control position for operating the main engine.

(10) Automatic Emergency Shutdown operates irrespective of control position. The Automatic Emergency Shutdown is provided in the event of overspeed, or when the engines exceed permissible limits for a pre-determined period. Once the Automatic Emergency Shutdown is activated, the EMERGENCY SHUTDOWN alarm sounds, and the indicator lamp illuminates on the main engine operating panel. The Automatic Emergency Shutdown operates irrespective of the control position.

These are some of the most important points which deck officers must consider when operating the main engine from the bridge. Below are some additional points that deck officers should consider when operating the main engines from the bridge location:

(1) Constantly monitor the engine load in rough weather. When sailing in rough weather and heavy seas, engine load must be monitored constantly. Stern swell often causes the propeller to emerge out of the water for a few fractions of a second. This causes a reduction in resistance near the propeller which results in the propeller racing and a surge in load on the main engine.

(2) If the remote control fails, transfer control to the engine room immediately. If the remote-control operation on the bridge becomes inoperable or ineffective, the control position should be immediately transferred to the engine control room. Bear in mind, however, that control cannot be changed over if the starter air main valve is not in service mode, or the auxiliary blower is not in auto position.

(3) If the main engine stops, move the manoeuvring handle to the stop position. If the main engine stops due to emergency shutdown or an automatic shutdown device, always move the manoeuvring handle on the control to the STOP position. In the event of a failure of the remote-control system, manual and automatic shutdown devices are provided to enable the engines to be shut down immediately to prevent any structural damage from occurring. To reset the emergency shutdown, the telegraph transmitter on the bridge must be put in the STOP position. In case of an automatic emergency shutdown, it must be borne in mind that the main engine is automatically shut down by a fuel cut-off due to overspeed, low lube oil pressures, or high temperatures of the thrust pad. To reset an emergency shutdown, the telegraph transmitter on the bridge must first be put into the STOP position and then the Emergency Stop button pressed in the control room. This will reset the system.

(4) Auto emergency slow down alarm can be cancelled to keep engine running. The AUTO EMERGENCY SLOW DOWN ALARM for the main engine functions like the AUTO EMERGENCY SHUTDOWN and can be

cancelled if required. This keeps the main engines running as per the telegraph order and does not slow the engines down.

(5) Check the starter air valve or auxiliary blower position in case of alarm while changing over the engine controls to the bridge from the engine control room. If any difficulty or alarm is observed when changing over the engine controls from the engine control room to the bridge, it is most likely due to the positioning of the starter air main valve or the auxiliary blower being in any other position other than the AUTO position.

(6) The imperfect bridge control condition alarm is mainly due to inappropriate positioning of the starter air main valve or auxiliary blower. If the alarm IMPERFECT BRIDGE CONTROL CONDITIONS activates, it will most likely be caused by the positioning of the starter air main valve or auxiliary blower in any other position other than AUTO or by abnormal bridge control air pressure. While stopping and starting the main engine, it must always be kept in mind that only a limited number of starts can be made with the charged air bottles for starter air.

(7) Follow all steps when removing the starting interlock for the main engine fuel oil supply. When removing the starting interlock to restore the main engine fuel oil supply, the turning gear should be disengaged and the emergency shutdown should be into the operational state. This is a safety feature that is provided in the main engine to prevent any operational mistakes when starting the main engine.

(8) Missed ignition restart takes place automatically. When the main engine keeps running for a pre-set time between the lower and upper limits of the critical speed range, the CRITICAL SPEED RANGE alarm will illuminate. The term MISS IGNITION is used when the main engine speed drops lower than the ignition level speed after the main engine is changed into fuel running. In this instance, the miss ignition restart will automatically occur after which the REPEATED START indicator lamp will light up once the engines have restarted.

(9) Exercise caution if the main engine is set to a limited revolution value. If the main engine is set to a limited revolution value by a ten key on the engine control room control panel, then the engine can be accelerated no more than the upper limit of the set revolution in case of bridge control. In such situations, if the telegraph transmitter on the bridge is moved to more than the position for limited revolution, the indicator LIMITED SPEED will illuminate. Whenever the starting operation is interrupted, the START IMPOSSIBLE alarm is activated. This can be reset by adjusting the telegraph transmitter to the STOP position.

Crash astern

In a crash astern condition, the ship's main engine is running at an ahead position. The telegraph transmitter is put into an astern position. The fuel is cut off due to difference between the running direction of the main engine and the telegraph transmitter. As the speed drops to a reversing level, the braking air is supplied.

The main engine is then reversed and accelerated to an astern ignition level. The starting fuel supply is started, and the starter air is cut off. Eventually the engine speed matches the telegraph transmitter position. Controlling the main engine from the bridge requires several important aspects and technical features to be accounted for by the ship's deck officers. The above-mentioned points are not exhaustive and must only be taken as reference when handling the navigational controls on the ship's bridge.

In this chapter we have covered some of the main duties of the OOW when taking over the navigational watch of a ship at sea, and the duties of the OOW during their watch period. Needless to say, this chapter has not covered every possible duty as many will be specific to the type of vessel, her area of operation, prevailing weather and sea conditions and the type of cargo carried on board. That said, this chapter should provide sufficient understanding of the nature of the navigational watch and the myriad duties and responsibilities the OOW has. In the next chapter, we will turn our attention to the Global Maritime Distress and Safety System (or GMDSS).

10 Global Maritime Distress and Safety System (GMDSS)

Throughout the late eighteenth and early nineteenth centuries, ships in distress depended on Morse Code to communicate their position to other passing ships. Though advanced for its time, Morse Code was far from perfect. Signals were often ignored, misinterpreted, or lost in the ether. Following the sinking of the RMS *Titanic* in 1912, the international maritime community realised a new method was needed to for ships to communicate their distress. This led to the development of the GMDSS. GMDSS was fully adopted by the IMO under SOLAS Chapter 4 on 1 February 1999. GMDSS sets a universal standard for communication protocols, procedures, and equipment that ships must use to communicate in times of distress. Under the GMDSS regulations, all cargo ships and passenger ships above 300 gross tonnes, and sailing in international waters, must carry GMDSS equipment on board. When the ship is in distress, the GMDSS transmitter on the ship sends a signal via satellite or radio wave which is then picked up by other GMDSS receivers. GMDSS is also used for sending and receiving MSI.

As the Earth is broadly spherical, the GMDSS signals are transmitted using different frequency bands depending on the location or 'area' of the vessel. This means for a signal to be sent and received, the GMDSS transmitter must be set to the correct area. When a ship uses GMDSS, it basically sends a distress signal via a satellite or radio communication equipment. It's also used as a medium for sending or receiving maritime safety information and as a general communication channel. In the GMDSS framework, there are different sea areas to allot the working equipment in the respective area. They are as shown in Table 10.1.

Table 10.1 GMDSS areas, ranges and equipment

Area	Range	Equipment
A1	20 to 50 nm	VHF DSC
A2	51 to 400 nm	VHF + MF
A3	70° N to 70° S	VHF + MF + One Inmarsat
A4	Above 70° N or below 70° S	HF + MF + VHF

DOI: 10.1201/9781003291534-10

To further understand Table 10.1, the following are the ranges regarding the frequencies in a specific band:

Table 10.2 GMDSS frequencies

Medium Frequency	MF	300 KHz to 3 MHz
High Frequency	HF	3 MHz to 30 MHz
Very High Frequency	VHF	30 MHz to 300 MHz

Types of GMDSS communications equipment

As per the GMDSS regulations, ships over a certain tonnage must carry specific types of GMDSS equipment on board. These include VHF radio, Inmarsat, NAVTEX and DSC. VHF is a set of radio frequencies that ships can use to communicate distress and send and receive MSI. The VHF range for distress signals is between 156 MHz and 174 MHz. Channel 16, which is set at 156.800 MHz, is specifically set aside for Distress, Urgency and Safety Communications. Channel 70, set at 156.525 MHz, is used for routine VHF DSC communications. GUARD channels are set above and below Channel 16 to avoid cross-channel interference. The GUARD channel frequencies are between 156.775 MHz and 156.825 MHz. The VHF set runs off a 24-volt DC power supply with J3E-type transmission for radiotelephony and G2B type for the transmission of VHF DSC comms. The International Maritime Satellite (or Inmarsat) is a system that connects ship-to-earth-to-station terminals through Inmarsat B, C and F77. The system provides telex, telephone and data transfer services between ship-to-ship, ship-to-shore, and shore-to-ship as well as priority telex and SAR telephony. NAVTEX is an internationally adopted automated system which is used to distribute MSI as well as weather forecasts and weather warnings, navigational warnings, SAR notices and other relevant safety information. DSC is a calling service that connects ships-to-ships, ships-to-shore, or shore-to-ships. The service transmits safety and distress information, usually on a high or medium frequency or VHF maritime radio.

Ships are legally required to have on board certain documents relating to GMDSS. These include the ship's radio licence, the radio operator's licence, the safety radio certificate, a GMDSS radio logbook, the International Telecomm unications Union (ITU) List of Cell Signs and Numerical Identities of Stations used by Maritime Satellite Services, the ITU List of Coastal Stations, the ITU List of Ship Stations, the ITU List of Radio Determination and Special Service Stations, an antenna rigging plan, and a valid shore-based maintenance certificate. The operation of GMDSS equipment requires specialised training as well as licensing from an approved maritime authority. For deck officers to be authorised to operate GMDSS, they must hold a General Operator's Certificate (GOC). To obtain the GOC, the officer must attend a compulsory GMDSS Operator course. This is generally aimed at deck officer cadets and consists of both written and oral examinations. On passing both the paper and oral assessment, the deck officer cadet is awarded the GOC and is authorised to use GMDSS equipment. The GMDSS Endorsement Certificate is a legal document that permits the certificate holder to work on foreign-flagged ships. Obtaining a GMDSS endorsement certificate qualifies the holder to operate specialised emergency radio equipment, including DSC, NAVTEX and the Search and Rescue Transponder (SART), take command of emergency situations on board in place of the master or other senior officer and to work on board foreign-flagged ships.

The GMDSS Endorsement Certificate must be revalidated every five years. The revalidation process involves the re-evaluation and renewal of the original GMDSS certificate. This is necessary to maintain the authenticity of the original GMDSS certificate. Though different maritime authorities set their own criteria, the universally accepted conditions for GMDSS Endorsement Certificate revalidation include performance of radio operations on board a seagoing ship that is fully fitted with GMDSS equipment for a period of at least 12 months within a preceding five-year period or at least three months within the preceding six-month period; successful completion of an approved training or GMDSS revalidation course provided by a recognised training provider within 12 months of sending the application revalidation; and completion of revalidation training every five years.

The criteria for transmitting GMDSS distress signals are set out in SOLAS and includes the following mandatory provisions:

(1) *Ship to Shore.* Every seagoing ship must have at least two separate transmission and receiver methods for sending ship-to-shore distress communications using either Emergency Position Indicating Radio Beacon (EPIRB), DSC or Inmarsat C;

(2) *Shore to Ship.* Every seagoing ship must be capable of receiving shore-to-ship warnings and distress alerts by DSC and NAVTEX;

(3) *Ship to Ship.* Every seagoing ship must be capable of transmitting and receiving distress signals ship to ship by VHF Channel 13 and DSC;

(4) *Search and Rescue Coordination Communications.* Every seagoing ship must be capable of transmitting and receiving SAR Coordination Communications by NAVTEX, HF, MF, VHF, and Inmarsat;

(5) *On Scene Communications.* Every seagoing ship must meet the required standard to coordinate SAR and other distress communications between ships at the scene of the incident through either HF, MF, or VHF;

(6) *Location Transmissions.* Every seagoing ship must be equipped with approved facilities for responding to maritime distress operations such as radar in accordance with SOLAS Chapter 5;

(7) *Transmitting and Receiving MSI.* Every seagoing ship must be capable of receiving MSI such as navigation warnings, chart corrections, weather forecasts and distress alerts through NAVTEX and DSC;

(8) *Shore-Based Networks.* Every seagoing ship must be fitted with general communications equipment for official, business, personal and private crew communications. This is usually provided through DSC and Inmarsat;

(9) *Bridge to Bridge.* Every seagoing ship must be capable of transmitting and receiving bridge-to-bridge communications. This is usually required in port or during pilotage using VHF for normal ranges and HF, MF, and Inmarsat for other ranges.

SAR signalling equipment

Modern SAR signalling equipment also uses radio waves and satellite signals to help rescuers locate distressed ships, or, in the worst cases, to locate lifeboats and life rafts and MOBs. There is a veritable selection of SAR signalling equipment available on the market and most ships will carry a variety of some, if not all of them. The most frequently used SAR signalling equipment includes the EPIRB, SART, and the Portable Marine Radio (PMR). The EPIRB is a type of emergency locator beacon that, upon activation, transmits a continuous radio signal that can be picked up and triangulated by SAR teams. There are three types of EPIRB currently in operation: (1) COSPAS-SARSAT. EPIRBs that function as part of the COSPAS-SARSAT system work on the 406.025 MHz and 121.5 MHz frequencies and are applicable for all sea areas; (2) Inmarsat E. INMARSAT E devices work off the 1.6 GHz band and may be deployed in sea areas A1, A2 and A3; and (3) VHF channel 70. This works on the 156.525 MHz band and is deployable only in sea area A1.

The EPIRB device contains two radio transmitters: one 5 watt and one 0.25 watt. Each operates at 406 MHz, which is the standard international frequency for distress signalling. The 5-watt radio transmitter is synchronised with a GOES weather satellite, which orbits the Earth in a geosynchronous pattern. When deployed, the EPIRB transmits a signal to the satellite. The signal consists of a digitally encrypted identification number that holds information such as the ship registration number or identifier, the date of the incident, the nature of the distress and the position of the EPIRB. A unique identifier number (UIN), or Hex ID, is programmed into each beacon during the manufacturing process. The UIN consists of 15 digits comprising numbers and letters and provides the unique identity of each beacon. UINs can be found on the external casing of each beacon – usually on a white label. The Local User Terminal (LUT) is a land-based satellite receiving unit or ground station that calculates the position of the beacon using Doppler Shift. Doppler Shift is the change in frequency or wavelength of a radio wave relative to the receiver and the transmitter. The LUT sends the beacon data to the closest regional Maritime Rescue Coordination Centre (MRCC). From here, the MRCC coordinates the rescue operation. In the event the EPIRB is not compatible with GPS signal-receiving equipment, the geosynchronous satellite orbiting the Earth will only pick up the signals emitted by the radio. This means the location of the transmitter, or the identity of the owner, will not be transmitted and received. This is because the satellite will only pick up trace elements of the radio signal, reducing the amount of information the receiver can interpret. As a result, the LUT will receive only a rough location of the distressed ship rather than an exact fix.

In practical terms, this means SAR crews can use the vague signal to plot an approximate location of the beacon, and gradually close in as the signal increases in strength, usually at around three miles. If an emitter transmits signals at 121.5 MHz SAR crews can pick up relatively precise locations at up to 15 miles. Accuracy can be improved if the beacon is fitted with a GPS receiver. To function, the EPIRB must be activated to emit signals. This is done by pushing a button on the unit; alternatively, some devices activate automatically when in contact with water. Water-activated EPIRBs are referred to as hydrostatic EPIRBs and are generally considered the best type as they do not require manual activation. That said, both hydrostatic and manually activated EPIRBs will only work when dislodged from their permanent brackets. This is achieved physically by a member of the crew or when immersed in water. All EPIRBs are battery powered, which is essential as power is most commonly the first utility to be affected during an emergency. The batteries are 12v and provide up to 48 hours of transmitting capacity. Battery expiry date checks should be carried out as part of the regular safety inspections. Most EPIRBs having an operational lifespan of between two and five years.

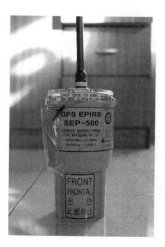 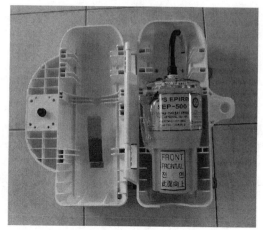

Figure 10.1 Example of a typical EPIRB.

It is possible for EPIRBs to activate accidentally. When an accidental activation has occurred, it is imperative that the false transmission is cancelled immediately by contacting the nearest coastal station or MRCC since otherwise an unnecessary SAR operation will be initiated. EPIRBs should be tested once a month to ensure they are fully operational. The procedure is simple:

(1) Press and release the test button on the EPIRB;
(2) Wait for the red lamp to flash once;
(3) Within 30 seconds, the red light and strobe should flash several times; and
(4) After 60 seconds, the device should automatically switch off.

EPIRB maintenance is necessary and straightforward and should be completed as part of the scheduled health and safety maintenance programme. The EPIRB must be inspected for visual signs of damage or impairment such as cracks and chips. Once inspected, the device should be wiped clean with a dry cloth to remove any residues such as salt. When wiping the device clean, carefully inspect the buttons and switches. The lanyard must be neatly packed into the container with no loose threads. The battery expiry date should also be checked regularly. Once an EPIRB has been activated, it must never be reactivated. Always return the device to an approved servicing agent for safe disposal.

The second type of SAR signalling equipment carried on board all ships is the SART. This is a self-contained, waterproof transponder designed specifically for maritime emergencies. There are two types of SART: radar-SART and GPS-based AIS-SART (automatic identification system SART). The radar-SART is used to locate survival craft or distressed vessels by creating a series of dots on the rescuing ship's radar display. The SART will only respond to a 9 GHz X-band (3 cm wavelength) radar and will not be seen on S-band (10 cm) or any other radar. The radar-SART may be triggered by any X-band radar within a range of approximately eight nautical miles (15 km). Each radar pulse received causes the SART to transmit a response which is swept repetitively across the complete radar frequency band. When interrogated, it sweeps rapidly (at 0.4 microseconds) through the band before beginning a relatively slow sweep (at 7.5 microseconds) through the band back to the starting frequency. This process is repeated for a total of 12 complete cycles. At some point in each sweep, the radar-SART frequency will match the interrogating radar, being within the pass band of the radar receiver. If the radar-SART is within range, the frequency match during each of the 12 slow sweeps will produce a response on the radar display. This line of 12 dots is equally spaced by about 0.64 nautical miles.

When the range to the radar-SART is reduced to about one nautical mile, the radar display may also show the 12 responses generated during the fast sweeps. These additional dot responses, which are also equally spaced by 0.64 nm, will be interspersed with the original line of 12 dots. These appear stronger and larger the closer the interrogating radar gets, slowly becoming arcs at first until the SART is within one nautical mile, at which point the arcs will become full circles, indicating the active SART is in the general area.

The general self-test procedure for the SART is simple:

(1) Switch the setting to TEST mode;
(2) Hold the SART in view of the radar antenna;
(3) Check that the visual indicator light is operational;
(4) Check the audible beeper is operational;
(5) Observe the radar display and identify the presence of concentric circles on the PPI;
(6) Check the battery expiry date.

Should the SART accidentally activate in live mode, switch the device off immediately and transmit a DSC alert cancellation message on VHF channel 70 and also

VHF channel 16 to all coastal stations, indicating the identification and position of the vessel. AIS-SART is a self-contained radio device used to locate survival craft or distressed ships. It works by sending automatic updated positions using an Automatic Identification System Class-A position report. The position and time synchronisation of the AIS-SART is derived from an inbuilt GNSS receiver (e.g., GPS). The PMR is a critical component of GMDSS and is located on the bridge. Being lightweight and durable, it consists of a portable handheld radio that can be used to communicate from the rescue craft. The IMO has mandated that PMRs must: be easily operated by unskilled personnel; be able to transmit and receive at 156.8 MHz (Channel 16) and 156.3 MHz (Channel 6); be able to withstand a drop of one metre onto a hard surface; be watertight to a depth of one metre for a minimum period of five minutes; and have a minimum power rating of 0.25 watts complete with a power reduction switch.

Figure 10.2 Example of a typical AIS-SART.

Furthermore, the antenna must be omnidirectional and vertically polarised, and the battery must have a power capacity of eight hours. Personal Location Beacons (or PLBs) are essentially EPIRBs but for individuals. PLBs are used to indicate distress when out of range of the emergency services. PLBs work in the same manner as EPRIBs and transmit on the COSPAS-SARSAT satellite system in the 406.025 MHz band range. PLBs are much smaller than EPIRBs and are designed to be carried on the person, whether at sea or on land. Once activated, PLBs transmit for approximately 24 hours. Pyrotechnic distress signals are a special type of distress signal that uses a self-contained and self-sustained exothermic chemical reaction for producing heat, light, gas, smoke, and sound. Ships use pyrotechnic distress signals to alert other ships of their distress. The use of pyrotechnic distress signals is covered at Annex 4 of the COLREGS and in SOLAS. The location of pyrotechnic distress signals and the minimum quantities to be carried on board are set out in both the COLREGS and SOLAS. Accordingly, on the bridge there must be a minimum of 12 rocket parachute flares, six handheld flares, four buoyant smoke signals with two on the port and two on the starboard sides and one line-throwing appliance. In each lifeboat, there must be a minimum of six handheld flares, four rocket parachute flares and two buoyant smoke signals. Each type of pyrotechnic distress signal has attributes which must be taken into consideration when deploying them in an emergency.

Handheld flares are a small cylindrical stick which, when activated, produce an intense red smoke or light without an explosion. The flare should be held out leeward when activated. Handheld flares can be used during both daylight and night-time hours. Rocket parachute flares are designed to fire a single red star to a height of approximately 300 metres (984 ft). When launched, the flare self-activates to produce an intense red smoke. At the peak of ascent, a parachute opens which reduces the rate of descent. This increases the time that the flare can be seen. Buoyant smoke signals are stored in compact floating containers. Used mainly during daylight, the buoyant smoke signal cloaks the position of distress with a bright orange smoke. This can also be used to aid rescue helicopters by indicating prevailing wind directions. Though not strictly a pyrotechnic distress signal, the line-throwing appliance is a counteracting device that can be used in emergency situations. When fired, it creates a connection or bridge between the distressed ship and the rescuing ship, upon which towing lines can be attached. The SOLAS regulations sets out the maintenance and disposal criteria for pyrotechnic distress signals. The general provisions for pyrotechnical distress signals are:

(1) Pyrotechnics must be kept in a lockable watertight storage container
(2) These containers should be kept in a dry, but accessible location
(3) Flares must be kept away from fuel and other combustible materials
(4) Weekly maintenance routines must be carried out in accordance with the manufacturer's instructions, LSA maintenance schedules and company ISM procedures
(5) When pyrotechnic distress signals expire, they must be removed from service and quarantined pending later disposal
(6) Pyrotechnics must be disposed of in port and never discarded overboard.

The provision and use of handheld flares are contained in SOLAS Chapter 3 regulation 25, accordingly handheld flares must:

(1) Be contained in a watertight casing
(2) Have a self-ignition system
(3) Not cause discomfort to the person using the flare
(4) Not cause damage to the vessel or survival craft

(5) Illuminate with a bright red colour
(6) Contain a simple and easily understood instruction diagram on the outside casing.

The provision and use of rocket parachute flares are contained in SOLAS Chapter 3 regulation 26; accordingly, rocket parachute flares must comply with the same provisions set out in SOLAS Chapter 3, regulation 25; and must:

(1) Have a minimum vertical height, when fired, of 300 metres (984 ft)
(2) Contain a parachute that automatically deploys when the flare reaches the top of its ascent
(3) Burn for a period of not less than 40 seconds with a minimum luminous intensity of 30,000 candelas
(4) Have a minimum rate of descent of 5 metres (16 feet) per second; and
(5) Have such an arrangement that the flare will not damage or burn the parachute once activated.

The provision and use of buoyant smoke signals are contained in SOLAS Chapter 3 regulation 37; and, accordingly, must:

(1) Be contained in a water-resistant buoyant container with a clear diagram for operation
(2) Not ignite or explode if used in accordance with the correct operational procedures
(3) Emit a high-visibility smoke with a uniform rate over a minimum period of three minutes
(4) Only emit smoke and not a flame when floating in calm water
(5) Emit smoke for a period not less than 10 seconds if immersed in water.

The provision and use of pyrotechnic line-throwing devices are contained in SOLAS Chapter 3 regulation 49; and, accordingly, must:

(1) Have good accuracy;
(2) Be contained in a water-resistant container with a clear diagram for operation;
(3) Contain a minimum of four projectiles, each with a line length of 230 metres (754 ft) in calm water; and
(4) Contain four lines with a SWL of not less than two kilos Newton.

Non-pyrotechnic distress signals

Non-pyrotechnic distress signals are the opposite of pyrotechnic signals in the sense that they do not involve an exothermic chemical reaction to operate. The most common forms of non-pyrotechnic distress signal equipment carried on board ships include EPIRBs, emergency radios, flying the NC flags in accordance with the ICS, operating the ship's horn, marker dyes, 'MAYDAY' or 'SOS', mirrors, orange signal flags, and slowly and repeatedly raising and lowering the arms. Other than the emergency distress telecommunications methods already discussed above, ships in distress can utilise other non-pyrotechnic means of getting the attention of passing ships and aircraft. The floating man, overboard pole or Dan buoy is a compact, self-contained device specially designed for maritime rescue and recovery. The Dan buoy is usually used during MOB incidents and is thrown overboard towards the casualty. The buoy is fitted with a yellow and red flag (ICS "O") and either a flashing lamp or strobe light. Prior to the adoption of modern distress technology, hoisting the national flag upside down could be used to indicate a ship in distress. This method is generally avoided today as it is easily misinterpreted as error and, in some cases, some national flags are the same irrespective of whether they are hoisted the right way up and upside down. As per the ICS regulations, the flag combination 'NC' indicates distress. Listed under Annex IV of the COLREGS, the Maritime Distress Signal Flag may be flown. This consists of a square orange cloth flag adorned with a black ball. Orange is used as it is the internationally recognised colour for distress.

If a Maritime Distress Signal Flag is unavailable, hoisting a length of orange cloth with a black square and a circle will also indicate a vessel is in distress. Also listed under Annex IV of the COLREGS are marker dyes, which have an approximate range of 50 metres (164 ft). Rarely used today, marine mirrors, better known as heliographs, may be used to attract the attention of other ships, or passing aircraft. The mirror works by reflecting sunlight in the direction of the approaching vessel. Mirrors not only reflect sunlight to pinpoint the position of the distressed ship or survival craft but are noncorrosive and never expire. They can be used for as long as the sun shines. The ship's foghorn may be sounded continuously to alert other vessels and shore authorities to the ship's distress. If all other methods have failed, slowly and repeatedly raising and lowering the arms, outstretched by the sides, can attract the attention of passing aircraft and small boats.

In this chapter, we have covered the main principles of the GMDSS system. As we might expect, GMDSS is a massive topic to which whole courses and books are devoted. Hopefully by the end of this chapter, you will have a better understanding of what GMDSS is and how critically important it is to modern-day sailing. In the next chapter, we will look at the procedures for laying anchor and the responsibilities of the deck officer when a ship is at anchorage.

11 Anchorage

Anchoring is one of the myriad operations that falls within the responsibility of the deck department. It involves the use of critical shipboard equipment and requires a high level of situational awareness. The key responsibility of the deck officer in charge (OiC) of the anchoring station is to use the anchoring machinery and available manpower for carrying out the operation safely and efficiently in accordance with the master's instructions. This chapter provides some practical suggestions for deck officers engaged in anchoring operations. First and foremost, for any successful operation on board a ship, is proper and efficient communication between the master, officers, and the crew members responsible for carrying out the task. In an anchoring situation, the bridge and the anchor station are the most important locations. The OiC should be thoroughly familiarised with the necessary reporting procedures. When anchoring, giving clear and succinct commands to the crew is as significant as receiving commands from the master. It is essential to regularly update the bridge on the status of the anchoring operation. There are two distinct anchor operations:

(1) Letting go (i.e., dropping the anchor); and
(2) Heaving (taking up the anchor).

In either case, the OiC has several responsibilities and duties to execute when overseeing the anchor operation.

Preparations for anchoring

Before starting the anchoring operation, the OiC must be clear regarding:

(1) The anchor to be used (i.e., the port or starboard anchor)
(2) How many shackles are to be lowered; and
(3) The manner which the anchor should be lowered (i.e., letting go or walking on).

DOI: 10.1201/9781003291534-11

Once the officer receives the command to prepare for anchoring, they must check and confirm the following:

(1) The presence of crew members wearing personal protective equipment (PPE)
(2) Again, confirming which anchor is to be used for the operation (port or starboard)
(3) The anchor lashings and bow stopper are removed prior to commencing the operation
(4) When using hydraulic windlasses, ensuring the pumps are started prior to operation
(5) Checking the working of the windlass and its controls
(6) If bow thrusters are likely to be used during anchoring, the OiC must ensure the required ventilations are open
(7) The hoisting of the anchor day signal (ball) is ready for hoisting after completing the operation
(8) PMRs or walkie-talkies are checked and fully charged; and
(9) The ship's sides are clear of any obstructions.

Operating the windlass

Ordinarily, the operation of the windlass is done remotely from the controls. It is always preferred that the windlass operation is carried out by a deck officer, provided the controls are positioned near the ship's side or in such position that the officer can maintain a clear line of sight of the anchor and its chain while operating the windlass controls. If this is not possible, the OiC should assign the duty to a skilled seaman guided with clear instructions.

Figure 11.1 Ship's windlass and mooring ropes.

Visually checking the anchor and its chain

As the OiC is responsible for overseeing the anchor operation and for reporting the position and stay of the anchor and its chain, it is recommended they keep a visual check themselves. Any abnormalities or problems must be reported to the master immediately. If the safety of the crew is likely to be affected, the anchoring operation must cease immediately until further notice.

Shackle tracking

Tracking the number of shackles lowered is done by visually observing the kender shackle of the chain. The kender shackle is bigger in size and is usually marked in different colour patterns or numbers for improved visibility. On modern ships, the length of the chain below the hawse pipe is digitally displayed on the control panel; however, it is better to maintain a visual check than to rely solely on the electronic system. If the officer is operating the windlass, then a crew member can be assigned the duty. For reference: one shackle equals 27.5 metres (90 ft).

Reporting

Reporting is one of the vital duties of the deck OiC of the anchoring operation. The OiC acts as the master's eyes during anchoring; hence every status of the operation should be reported in good time to the master. The most important details to be reported to the master are the anchor position. When anchoring operations are underway, the anchor-chain position is a matter of high concern. Positions are reported according to the 24 hour clock format, with the bulb being 12 o'clock. Positions on the starboard side are reported, starting from 1 o'clock to 2 o'clock and so on; positions on the port side start from 11 o'clock to 10 o'clock, etc. Cardinal points are reported every 11.25 degrees; for example, two points on the starboard bow. Along with the position, the stay of the chain must also be reported. Stay of the chain is the tendency of its movement. When reporting the chain stay, the following terms are used:

(1) Short stay – when the chain is leading in a short range from the ship's side
(2) Medium stay – when the chain is leading in a medium range from the ship's side
(3) Long stay – when the chain is leading in a longer range from the ship's side extending from the hawse pipe; and
(4) Up and down – when the chain is vertically leading parallel to the ship's side. It will not extend and will be leading vertically downwards from the hawse pipe to the seabed.

Different officers will have different reporting styles. It is perfectly acceptable for individuals – who should be encouraged – to use a reporting method the individual is comfortable with, so long as the master and the rest of the anchoring team are clearly able to understand what is being reported. In normal situations, when letting go the anchor, the chain needs to be stretched out for the anchor to hold the vessel. After dropping the anchor, the chain stay will be longer. When the anchor is holding to the seabed and the chain is settling down, the stay comes gradually to medium and then to a short range. Finally, the chain will go up and down, which means the anchor is holding and the chain has settled.

Safety considerations during anchoring operations

The OiC of the anchoring operation is responsible for the safety of the equipment and the crew members involved. The officer must watch out for themselves and their crew. Unsafe practices must never be tolerated and wherever possible corrected before carrying on with the operation. The OiC should be able to lead the crew and guide them to carry out the operation safely and efficiently. With regards to safety, the following points should be noted:

(1) Always wear appropriate PPE, including safety helmet, gloves, goggles, and earmuffs (if required)
(2) When checking the anchor and its chain, find the best position to observe the anchor from on deck and avoid leaning over the ship's side
(3) When letting go the anchor, always maintain a safe and clear distance from the windlass
(4) When lowering or heaving the anchor on gear, do so at a gradual speed and avoid sudden and repeated alterations of windlass and its direction of movement.

Duties of the OOW when anchoring

Calculating the swinging circle

When a ship is anchored at sea, it is almost entirely left at the mercy of the prevailing currents and tides. This is because the ship has a single point of pivot. As currents and tides pass by the ship, they cause the vessel to move accordingly. This is referred to as swing. It is critical when deciding where to anchor to consider the vessel's swinging circle, or the diameter of the ship as it pivots around the anchor. Although the swinging circle will vary with changes in tidal levels, the weather, ballast, and various other factors, it is important to have a fair idea of the radius around which the ship might move. To determine the swinging circle, the following calculation is used:

$$\text{Swinging Circle (M)} = \text{Length Overall} + \text{Length of Cable} - \text{UKC}$$

Keeping a visual on traffic in the vicinity

When the ship is at anchor, it is ever more important to maintain a close visual on other ships and watercraft in the immediate vicinity. The greatest danger during anchorage is when another vessel starts dragging after anchoring. Dragging might occur due to changes in tidal levels, changes in weather or due to the brake giving away to a lot of yawing. The danger here is primarily to ensure that the vessel does not drag and to check that no other vessels are drifting towards your own vessel. Using the bow stopper when at anchor can help minimise the risk of dragging.

Keep a constant check on the ship's position

This point is generally laid down in the master's standing orders and almost always mentions the intervals at which the position must be plotted on the chart. Using the GPS, the radar, or both, helps establish the exact position of the ship, which in turn helps to determine whether the ship is stationary or dragging.

Display appropriate lights

Displaying the appropriate lights enables other vessels to know that you are anchored, thereby making it evident that they should not anchor too close to your vessel, and provides an indication of proximity should dragging occur.

Keep a constant VHF watch

If slated to receive a pilot on board or for information on vessels in the area, it is critically important to maintain a constant VHF watch. The coastal authorities generally promulgate useful information, such as weather reports and shipping activities, that might be integral to the safety of the vessel.

Alert nearby vessels when required

If another vessel seems to be coming too close to your own vessel, attempt to get their attention on VHF or by flashing the Aldis lamp. If either, or both, of these methods fail to use any available means to attract their attention.

Watchkeeping at anchorage might seem an easy task, for there is barely any movement. However, the very fact that the vessel is stopped dead in the water makes it vulnerable to a variety of hazards. As is always the case, the OOW must be vigilant and use the assistance of additional lookouts where necessary. In areas prone to piracy, the importance of alertness simply cannot be stressed enough. As the ship is stationary, it makes for a prime target. Diligence on the part of the OOW will go a long way in ensuring the vessel is kept safe.

Dragging anchor

A merchant vessel is said to be dragging anchor when the vessel drifts without holding power despite being anchored. Dragging anchor has been the cause for several accidents over the past few decades, including collisions, groundings, and stranding. There are many reasons why a vessel may anchor, including waiting for a berth prospectus, cargo discharge and loading; maintenance; or waiting to carry out instructions from the ship's owners or charterers. The main reason why a vessel will drag its anchor is due to rough weather conditions. In such situations, it is extremely important for the bridge team to collect as much information about the prevailing conditions as possible to put into effect contingency measures for preventing the dragging of anchor as much as possible. Some of the important parameters that need to be taken into consideration are:

(1) The prevailing weather conditions for the area
(2) Safe and ideally sheltered positions for anchoring the vessel
(3) Wind and tidal behaviours in the anchorage area; and
(4) The contact information for the local or nearest port and coastal authority. At most ports, it is inevitable for a vessel to have to wait at anchorage and the time at anchorage can be as much as days or even weeks. During such times, the master and ship's crew should identify any potential dangers to the ship and make all the necessary preparations.

A vessel dragging anchor is a threat to its own safety and to other vessels in the vicinity, often leading to emergency situations such as collision, grounding, or stranding, depending on the manoeuvrable condition of the ship. In such situations, a quick assessment can only be achieved by vigilant bridge watches, contingency planning to respond to an emergency, quick but measured responses, and good reasoned judgement. Always bear in mind it will take some time to weigh anchor and restore the vessel to its full manoeuvrable condition, therefore every possible effort must be made to ensure the vessel is anchored in plenty of sea room. Where a vessel is considered likely to drag anchor, there are some actions which the master can take to minimise the effect and reduce the risk to the vessel. Taking on heavy weather ballast, though accounting for vessel stability and the depth of water below the keel is one such example. Paying out more anchor cable depending on the size of the vessel and the weather conditions can help alleviate some of the stress load. Keeping a safe distance from other anchored ships and subsea structures such as pipelines. Leaving sufficient room for manoeuvring. Weighing anchor and shifting the vessel to an alternative position away from the vicinity of other vessels, provided permission is received from VTS, the port authorities and ship owner's orders. Increasing the efficiency of the bridge team by adding an extra lookout and finally by keeping the ship's main engines on standby.

How to assess whether the vessel is dragging anchor

When a ship is dragging anchor, it may not be immediately obvious, especially in heavy swells or rough seas. To assess whether the ship is shifting from its anchorage position requires checking the ship's position at frequent intervals. This will determine whether the vessel is outside the swinging circle. Use all available means – both visual and electronic – such as ECDIS, GPS and radar to make the assessment. If the vessel has deviated from its swinging circle, it is likely to be dragging its anchor. Whilst there is no alternative to keeping a vigilant watch when the ship is at anchorage, there are other tell-tale signs that may indicate the vessel is dragging. These typically include:

(1) The bow not being able to withstand the wind
(2) Checking the anchor chains for slippage – a small pole with a cloth can be tied to the links to measure any potential slipping of anchor chains
(3) Extra vibrations and weight on the anchor cable
(4) Variance in the SOG – when the vessel is swinging, the SOG can increase uncontrollably
(5) Checking the course recorder for a figure of eight motion locus; and
(6) Monitoring the position and distance of vessels nearby.

What actions are to be taken if the vessel has started dragging anchor?

If after assessing the condition of the vessel it is decided the ship has started dragging anchor, there are actions that must be carried out immediately. First and foremost, the master must be informed irrespective of the time of day or night. Inform the engine room and request the main engine is started to provide power to the windlass; this will need the permission of the master. Make the vessel ready for manoeuvring. Stop all cargo operations and prepare the vessel for manoeuvring. Let go any cargo barges and crane barges if alongside. Alert VTS and other vessels nearby about the condition and inform them about the actions to be taken. Seek permission for re-anchoring. Start heaving up the anchor and once the vessel's manoeuvrability is restored, shift the anchorage position where drifting can be safer or make way to open sea. Deploy more cables or drop a second anchor (this is not recommended for larger vessels) before the speed of dragging of the vessel increases. This can stop smaller vessels from dragging anchor at a very early stage before the ship is pressed to the leeward side with increasing speed. Use the bow thrusters, main engine, and steering gear to manoeuvre. Remember it becomes more difficult to weigh anchor when the vessel is pressed more to the leeward side; it also takes considerably more time. Use the bow thrusters to stem the wind. Do not override the anchor – especially in shallow waters – as the vessel may impact on the anchor during pitching. If the situation permits, let the vessel drag in a controlled manner. This is not recommended in areas where offshore work such as oil and gas operations are carried out, which can result in damaging submerged pipelines and cables. If weighing the anchor is not possible, and all other efforts have been exhausted, release the bitter end and let go the anchor completely. This must be the very last resort, as a ship without a minimum of two anchors is not considered seaworthy. If the weather permits, call for tugboat assistance and request to be taken to safe harbour.

Most accidents, collisions and groundings happen when the vessel is at anchor mainly because there is no early prediction of the anchor dragging. Time plays a vital role in areas of high vessel density and this time lapse often results in difficulty restoring the manoeuvrability of the vessel. Subsequently, always ensure that proper contingency plans are available and set in motion at the earliest possible indication.

In this chapter, we have looked at some of the main principles of anchorage and the duties of the OOW when a ship is preparing to conduct anchoring operations, during anchoring operations and whilst the vessel is at anchorage. In the next chapter, we will look at the role and responsibilities of the marine pilot.

12 Maritime pilot

Whilst the ship's master is in overall command of the vessel, the role and function of the maritime pilot is of equal, if temporary, importance. As the name suggests, the maritime pilot supports the master in bringing the ship safely into and out of port. Whereas the master is concerned with the navigation and handling of the ship, the pilot's role is to advise, and counsel, the master on the most effective and safest ways of manoeuvring the ship. The role and importance of the pilot increases exponentially as the size of the vessel increases and the manoeuvrability of the vessel decreases. Indeed, ships that carry hazardous cargoes or dangerous products, such as oil tankers, are more dependent than most on the pilot's expertise. The profession of the maritime pilot requires years of training and expertise about local waters, tides, currents, and weather conditions. If the entry to the port is narrow or has unusual hazards such as sandbanks or precautionary channels, then a pilot must be brought on board to provide expert guidance. The pilot must therefore work alongside the ship's master, complements the expertise of the master about their ship with expert knowledge of the port and its seaways.

The duties of the maritime pilot are similar the world over though individual countries and port authorities may exercise different laws and regulations governing the conduct of pilots and the division of responsibilities between the pilot, the master, and the ship's officers and crew. For example, in English law, under section 742 of the Merchant Shipping Act 1894, a pilot is defined as "any person not belonging to a ship who has the conduct thereof". In other words, the act defines a pilot as someone other than a member of the ship's crew who has control over the speed, direction, and movement of the ship. That said, the express duties and responsibilities of marine pilots were not codified until the passing of the *Pilotage Act 1987*, which sought to govern the management of maritime pilots and pilotage in British harbours. Because of the responsibilities exercised by the pilot, maritime pilots are required to have in-service experience prior to qualifying for a pilot's licence. This includes having extensive local knowledge of the port and its surrounding areas. Again, specific qualifications and experience differs greatly from one country to another, and even from region or port to another. For example, in the United States, each state is responsible for setting their own criteria. The California Board of Pilot Commissioners, for instance, requires trainee pilots to have, as a minimum: a master's licence, at least two years' command experience on

DOI: 10.1201/9781003291534-12

tugboats or deep draught vessels, as well as passing a written examination and simulator exercise, followed by up to three years' training to gain experience on a variety of different vessel types and docking facilities. Even when qualified, the pilot must continue their training to maintain professional competence.

Role and function of the maritime pilot

Typically, the pilot joins an incoming ship before entering the port's waters. This is referred to as the *pilot boarding area*. The pilot may be brought on board by helicopter or pilot boat. With the latter, the pilot may be required to climb a vertical pilot ladder as high as 12 metres (40 ft) in heavy swells and strong winds. Add to this the fact that both the pilot boat and the boarding vessel will be moving, this is an extremely dangerous undertaking that requires a stern constitution. When a vessel is leaving port, the pilot usually boards the ship before departing and is returned ashore once the vessel has successfully left the port's confines. Because many ports and harbours have narrow navigation channels, hidden hazards, and other points that the master may not know about, it is often a legal requirement for ships entering port to engage the services of the pilot. The pilot may or may not request the support of tugboats. Despite the pilot being the expert of the local waters, legally, it is the master who retains full responsibility for the safe navigation of their vessel, even when the pilot is on board. If the master feels there are clear grounds to disregard the advice of the pilot, the master has the authority to relieve the pilot of their duties and request a new pilot, or, if not required, to navigate the vessel without one. Irrespective of this, the pilot remains under the master's authority despite being outside the ship's structure of command. This means the pilot must do as the master orders (within legal limits) but is under no obligation to follow the commands of any other of the ship's officers. The only exception to this rule[1] is the Panama Canal, where the pilot has full responsibility for the safe navigation of the vessel. In some maritime domains, it may be acceptable for the ship's deck officers to exercise the role of the pilot. This is predominantly the case where the deck officers have strong local knowledge and experience of navigating into and out of port. This usually encompasses regular ferry services such as the cross-channel ferries between England and France. In these circumstances, the vessel may be issued with a *Pilotage Exemption Certificate*, which relieves the master of the need to take a pilot on board.

Preparing to receive the pilot on board

Maintain communications between the vessel, VTS and the pilot

Before the pilot is authorised to board the vessel, the OOW must first carry out several core duties. These include:

(1) Collecting the necessary information, such as the VHF channel used by the pilot and the position of the pilot grounds and other services. This information can be obtained from *Nautical Publication NP 286 ALRS* (Admiralty List

of Radio Signals) for the specific port. VTS will also provide this information if requested

(2) Maintain watch on the appropriate VHF channel used by the pilot. This is mostly a dedicated channel that is separate from the channels used by VTS and Port Control

(3) Calculate the ETA of the vessel to the pilot boarding ground; observe the prevailing weather conditions and sea states. These should be reported to the master and VTS

(4) Contact the pilot on VHF and provide the relevant information (such as ETA) to the pilot ground, the vessel's draught and freeboard of the vessel, the length of the vessel, type of propeller, the vessel's current speed over ground and course made good. This is an opportune moment to receive information such as the berthing prospectus, the side of the vessel that the pilot ladder is to be rigged, the required height of the pilot ladder from water level, and the speed and course to be maintained. This information must be recorded and reported direct to the master

(5) Monitor the pilot boat whilst maintaining clear communications between the pilot boat and own ship. The officer in charge of the pilot boarding should be present near the point of access to receive the pilot on board and later when he disembarks. The pilot on board and disembarkation time should be noted and recorded.

Figure 12.1 Pilot boat outside Öja island (Landsort), Stockholm archipelago's most southern point.

Preparing for the pilot to board

When preparing to the receive the pilot on board, there are various tasks and checks the master and OOW must complete to ensure the period of pilotage is carried out efficiently, including:

(1) *Liaise with VTS.* It is important to establish the correct ETA of the pilot with VTS on VHF. Quite commonly, there will be several vessels all eagerly waiting for their pilot, and delays are not uncommon. Keeping in regular contact over VHF with VTS will ensure the vessel is kept up to date with the latest pilot information. It is important to ensure a log of all communications between the vessel and VTS are kept

(2) *Rig the pilot ladder.* VTS will generally inform the bridge on which side the pilot will be boarding and the requirement of height above water. For this reason, the pilot ladder must be rigged observing strict standards of safety. If requested, the gangway may also need to be rigged accordingly. Both the ladder and the gangway (if applicable) should be arranged as per the master's instructions, with the appropriate safety measures provided (such as life jackets, life buoys, handheld VHF radios, etc.)

(3) *Check the required equipment and systems are operational.* It is always a good idea to synchronise clocks and watches. Keep both radars switched on and ready. Set the echo sounder to constantly monitor the ukc. Adjust the squelch and volume controls of the VHF. Ready the appropriate large-scale charts and put in order of anticipated use

(4) *Prepare the master/pilot exchange forms.* These are important legal documents that must be completed, countersigned, and stored on board

(5) *Inform the engine control room (ECR) of the pilot's estimated arrival time.* Prior to the pilot's arrival, the engine room should be thoroughly briefed and prepared for the pending manoeuvres and all systems and machinery on standby

(6) *Unless necessary, suspend all deck work and assign crew members specific tasks supporting the pilotage.* At this point the mooring crews should be checking and readying the mooring equipment and ropes. All crew members should be instructed to stand easy in preparation of making way

(7) *Engage the manual steering prior to the arrival of the pilot.* The steering should be changed over from automatic to manual to allow the helmsman time to accustom themselves to the ship's movement

(8) *Hoist the pilot flag.* The last action to be taken in preparation for receiving the pilot on board is the pilot flag. This advises all other vessels in the vicinity that the ship is preparing to enter port under the tutelage of the pilot.

Rigging the pilot ladder

The pilot ladder is a special type of rope ladder used on board ships for the embarkation and disembarkation of maritime pilots. It is one of the first things the pilot will notice about the ship and how the master regards himself and his crew. A sloppy pilot ladder rig indicates a sloppy onboard attitude. To prevent this, it is important the pilot ladder is well maintained and rigged properly. Even the slightest oversight can lead to fatal accidents.

Real Life Incident

On a small cargo ship, the pilot was preparing to disembark with a good lee The bottom rung of the pilot ladder was slightly above, and clear of, the deck of the launch. The pilot inspected the ladder as best he could from the main deck to ensure it was properly rigged, secured, and free from any obvious defects. All appeared correct. But, as soon as the pilot committed his full weight to the ladder, the ropes on each side of the ladder parted simultaneously at the point where they went over the rounded fishplate at the sheer strake. The pilot fell two metres (six feet) to the deck of the pilot launch, spraining his right foot and ankle, which took the impact of the fall. His injuries might have been much worse had the freeboard of the vessel been higher. On investigation, it was found the root cause of the accident was failure of the pilot ladder due to poor maintenance.

The pilot ladder should be capable of covering the entire length from the point of access to the water level. The height of the water level to the top of the ladder is contingent on the height of the pilot boat. This information is provided to the bridge by VTS. To prevent accidents like the one described above from happening, there are several key points to be noted when rigging the pilot ladder:

(1) The top portion or head of the pilot ladder should be secured at the strongest point available on the vessel

(2) The pilot ladder should be positioned and secured in such a way that it is clear from any discharges from the ship, with the parallel body length of the ship, and as far as practicable within the halfway length (midship) of the ship

(3) All steps on the pilot ladder should rest firmly against the ship's side. On certain vessels, where constructional features such as fenders or rubbing bands prevent the ladder from resting flush against the ship's side, special arrangements must be implemented instead to facilitate the safe embarkation and disembarkation of the pilot

(4) Two-man ropes, between 28 mm and 32 mm in diameter and made from manila rope or some other material which provides a firm grip for climbing the ladder, should be rigged along the side of the pilot ladder where requested.

During night boardings, the whole length of the pilot ladder, point access, and egress point must be well illuminated. A life buoy with self-igniting light and a heaving line should be kept at the ready. If requested, handheld stanchions and bulwark ladders should be made available. If the point of access from sea level is more than nine metres (29 ft), a combination ladder should be used instead. A combination ladder is a combined pilot ladder and accommodation ladder. This is a common arrangement found on vessels with high freeboards. The accommodation ladder is rigged in such a way that it leads aft of the vessel and has a slope angle not exceeding 55 degrees.

Figure 12.2 Pilot boat, Bay of Sevastopol.

Maintaining the pilot ladder

Pilot ladders must be regularly inspected for wear and tear, missing wedges, and damage to the steps. The ladder steps should never be painted and must be kept clean, and free from oil and grease. The steps should be equally spaced between the side ropes, and the distance between two steps should be uniform, and horizontal. Any damaged or defective ropes and steps must be replaced immediately. The side ropes are made of manila rope. They should be continuous and free from ties and joints below the first step of the pilot ladder. The shackles used to secure the pilot ladder should have the same strength and durability as the side ropes. Once the pilotage operation is over the pilot ladder should be secured instead of left hanging on the ship's side. The pilot ladder should be stowed in a dry and well-ventilated space, clear of the deck, and fitted with a cover to conceal the ladder from sunlight, chemicals, and paint spills. Furthermore, the pilot ladder is used exclusively for the

Figure 12.3 Dutch pilot boat, *Origin*, Port of Ijmuiden, Netherlands.

boarding and disembarking of the pilot and must never be used for any other purpose, including general maintenance or working aloft. Full guidance on the design, construction, use and maintenance of pilot ladders is contained in SOLAS Chapter V under the regulations pertaining to the safety of navigation.

Once the pilot is on board

Once the pilot is on the bridge, the first thing they will seek is the *Pilot Card*. This document is held on the bridge and provides the pilot with the vital statistics and specifications of the vessel. The *Pilot Card* contains the vessel's LOA, its beam, deadweight, tonnage, draught forward and aft, the engine's RPM and rated speed (both ballast and loaded) during the different phases of ahead and astern,[2] the nature of the propeller (i.e., whether constant pitch propeller (CPP), variable pitch propeller (VPP), normal right-handed, Schottel, Voith Schneider, etc.), her bow and stern thruster's power (if available) and any other technical details relevant to the operation of the vessel. Once the pilot has absorbed the information contained on the *Pilot Card*, the pilot can determine the best course of action. The pilot will also need to confirm the canting of the vessel (i.e., in which direction the bow or stem will move once according to screw race and transverse thrust once the engines are put on astern) and will also review and confirm the state of the navigation and steering equipment on board the vessel. Once the pilot is content with the state of the vessel's systems, the process of pilotage can begin in earnest. At this point a very delicate balance begins to develop between the expertise of the master and his ship and the pilot and his waters. This balance is crucially built on a fragile foundation of trust and respect. As in all professional settings, it is easy to let personal differences and dislikes stand in the way of professional conduct, but where the safety and integrity of the ship, its cargo and crew are concerned, it is incumbent on both the master and the pilot to work together in synergy. With that being said, the master retains ultimate authority and has the discretion to override the pilot's suggestions, albeit at the master's own risk. Sometimes it may be necessary for the pilot to suggest a course of action that the master disagrees with. In these situations, the master may decide not to endorse the pilot's recommendation. Of course, it may well be that the master has an overly cautious nature, in which case the pilot should attempt to explain the reasoning behind the recommendation. Again, after doing so, the master may choose not to endorse the pilot. If in this situation, the master, by ignoring the pilot, endangers the ship or the port facilities, the pilot must inform the port authorities that he no longer has the "conn" of the vessel. Should the master wish to reinstate the "conn" of the vessel to the pilot, this must be given first orally and then in writing. It is in the master's and the pilot's best interest to work together, and to come to some accommodation that meets both requirements.

Duties of the OOW

During pilotage, the OOW assumes several duties which they must execute. Although the master is in direct command of the vessel, and the pilot will advise the master on the best courses of action, the OOW is still responsible for the safe conduct of the vessel throughout their watch. It is common practice for the pilot to instruct the helmsman about the intended course of action. Any specific operational requirement by the pilot, so long as the master approves, is to be followed without question. The OOW must remain vigilant and observe the helmsman as they carry out the pilot's and master's instructions. If during the pilotage the helmsman becomes fatigued, the OOW must ensure there is a suitable replacement on standby to relieve the outgoing helmsman.

Figure 12.4 Changing the pilot in the Suez Canal, Egypt.

Important navigational marks should be recorded in the Ship's Manoeuvring Book as they correspond to the navigation chart. The position of the vessel should be plotted on the chart as deemed necessary by the master. If this task is more than the OOW can handle by themselves, bearing in mind they must also carry out their other assigned watch duties, the OOW may request assistance from an Officer Cadet (if present on board) or senior member of the ship's deck crew. When the vessel is proceeding to berth, the OOW must follow the master's instructions regarding the ship's the manoeuvring speed and any other such orders as relayed by the pilot to the master.

Systems and tools used by the pilot

Today, piloting in restricted waters has become a highly specialised job aided by modern technology. Gadgets, instruments, and software tools are always in the process of evolving as newer advancements in navigational aids and technology emerge. The days of anchoring during mist, fog and blizzards are long gone. Also consigned to the footnotes of maritime history are the days when navigators had to take a position fix taking into consideration the external reference points across the navigating channels and plotting lines using a bearing indicator. The point of intersection of the lines on the chart used to give the position of the vessel. This point of intersection is called a *fix*. Today, GPS has assumed this task and provides a far more immediate and accurate position. Moreover, new advanced control systems, which offer increased accuracy and integrity to GPS, are constantly in development. Subsequently, maritime pilots today use several navigational tools and systems.

Electronic Nautical Charts (ENC)

Prior to the development of ECDIS and electronic nautical charts (ENC), navigators had to work in tandem with the pilot and depended solely upon paper charts. These charts would provide a two-dimensional view of the sea or riverbed and the local topography to assist in safe navigation. Charts also indicate navigational hazards, sudden elevations of the seabed, wrecks that block the navigation channel in restricted waters, all types of local manmade structures, the position of bridges, ports, structures on shore, the position of guiding buoys, turrets, obelisks, and other shore-based references. These charts were prepared by national hydrographic departments (for example, the UK Hydrographic Office) and were updated several years apart. This made navigation vulnerable to sudden changes on the seabed of the channel. Physical charts also required huge spaces to be laid out on a chart table, which inevitably blocked the chart room. To avoid these problems, ENCs were developed to move navigation from paper to digital.

There are two main types of ENCs: the raster chart and the vector chart. While the first is merely a scanned variety of the paper navigational charts discussed above, the second is more data-oriented. Though they are hidden, the data at a particular position are instantly available at the click of a mouse. The disclosure of important navigational data is achieved when the ENC is integrated with navigational software, such as the Electronic Chart Display Information System (or ECDIS). All electronic nautical charts conform to the guidelines set by the International Hydrographic Organisation. Moreover, these charts are regularly updated according to resolutions adopted by the IMO, which invited governments of member countries to conduct hydrographic surveys and publish and disseminate regular nautical information to aid safe navigation. Numerous chart-plotting software also exists, which can make navigation easier still. One such software platform is SEA CLARE, which is a computer-based software designed for the Microsoft Windows 2000® operating system and above. When connected to GPS, SEA CLARE shows the current position, speed and heading of the vessel in real time.

New charts can be fed automatically in text file and tracks can be saved for later viewing. Entries can be manually updated and entered. Numerous waypoints can be created to assist navigation. The only drawback to the SEA CLARE system is that is requires an advanced GPS transmission capability called NMEA 0183 (which is a specification or protocol developed by the National Marine Electronic Association, Maryland, USA).

Other bridge instruments which conform to this protocol can be connected. Data such as the depth of water below the keel, wind direction, the ship's gyro heading, and even AIS, can be connected to track nearby vessel movements. A simpler and less costly alternative to SEA CLARE is a software platform called PC PLOTTER, which is generally used by smaller craft such as yachts and fishing boats. This platform requires a low-cost Dual Channel Parallel AIS Receiver that can receive signals from large and small ships. As we have already discussed earlier, AIS is the vessel tracking system which is used to locate other ships in the vicinity by automatically exchanging data with nearby vessels and AIS base stations. Local VTMS (Vessel Traffic Management Systems)/VTS (Vessel Traffic Service) is offered by ports where traffic is busy, but AIS acts as an additional support tool. Another instrument that reduces the workload of the pilot is the marine ARPA-enhanced radar. Radar, as we know, is a radio detection and ranging device that reflects electromagnetic waves. Ships, aircrafts, buildings, motor vehicles, marshy lands, water bodies, and even low-lying clouds all reflect radio waves and hence are visible on the radar screen. The ARPA can calculate the speed of tracked vessels, the vessel's course and predict the closest point of approach. Furthermore, the relative speed between the vessel and a static point such as a land mass can be calculated, with the collision point and time clearly indicated.

Procedure for 'Pilot Away'

The procedure for *pilot away* is similar in many respects to the pilot boarding. As a matter of extending professional courtesy, the pilot is ushered out as he is ushered in. Once the forms are signed, and all other paperwork has been completed, the OOW should ensure the pilot disembarks safely. It is advisable to inform the boarding party in advance so they may have the pilot ladder or combination ladder rigged on the required side. As with boarding the vessel, all safety measures must be adhered to. The process of pilot embarking and disembarking is simple but extremely dangerous. It is risky for the crew who put themselves in harm's way when rigging the pilot ladder and gangway, which is especially so on older vessels. The OOW should be vigilant that all safe working practices are complied with. For Officer Cadets, pilotage provides a great experience. If possible, and deemed appropriate by the master, Officer Cadets ought to be included in the pilotage operation. As far as the senior members of the bridge team are concerned, however, pilotage is a hazardous operation that requires concentration and experience.

Whilst the pilot has expert knowledge of the waters, the master is the expert of their ship. Any errors committed by the pilot do not exonerate the master from their responsibilities. In that sense, the master is responsible for the pilot's actions.

If there is any doubt whatsoever in the master's mind, they have the right and authority to challenge the pilot. If after the pilot has explained the rationale for their proposed course of action the master remains uncertain, it is their responsibility – and theirs alone – whether or not to follow the pilot's instructions.

Notes

1 Correct as of September 2021.
2 For example, dead slow ahead.

13 Mooring and berthing

Ports are incredibly complex facilities that thrive on efficiency and well-organised supply chains. By ensuring there is no lag between vessels entering the port, having their cargo loaded or unloaded, refuelled, and inspected, cleared to depart, and finally pushed clear of the berth, ports are central facilitators to global trade and national and regional prosperity. Many ports today utilise the very latest in technology to minimise the time ships spend alongside, therein reducing turn-around times and maximising berth space. To put this into perspective, China has some of the most technologically advanced port facilities anywhere worldwide. In 2019 the Chinese port of Ningbo-Zhoushan handled some 1,120,090 tonnes of containerised cargo; in 2020, this increased 4.7%, to 1,1720,400 tonnes. Shanghai, by comparison, handled 716,770 tonnes in 2019 and 711,040 tonnes in 2020, a drop of 0.8%.[1] Part of the reason why Ningbo-Zhoushan and Shanghai can handle such vast quantities of cargo lies in the automation of cargo loading and discharge systems, specialised truck trailer docking sites, railway wagon identification mechanisms and advanced load-bearing cranes. When added together, these make for an extremely efficient, cost-effective and time-saving operation. Despite these quite amazing feats of technological advancement, port operators can do very little until the ship is safely alongside. This, unfortunately, remains an agonisingly slow operation.

Berthing plans

In the busiest ports, in an average month, as many as several hundred ships may berth. This constant coming and going requires constant planning, revisions, and replanning. This is where the *berthing plan* comes into play. Berthing plans are integral to the efficient allocation of port resources for all incoming and outgoing vessels. They enable the port facility to plan for each vessel at least one month in advance, though sometimes berthing plans can be drawn up to cover as much as six months in the future. In its most simplistic form, the berthing plan acts as a snapshot of each berth at any given time. By having a high-level overview of what ships are in and out, and what cargoes are being loaded and discharged, the port can look out for potential bottlenecks and other problems and initiate solutions in a timely manner. This means berthing plans must be, on the one hand, extremely detailed; on the other hand, however, they are capable of change without causing

DOI: 10.1201/9781003291534-13

too much aggravation to other port operations. To prepare the berthing plan, the port gathers information on every vessel expected to arrive and depart the port within a given timeframe. This information is provided every time a shipping company or charterer books a berth at the port on a certain day. The details are submitted to a central booking system. From there, the port operations team develop the berthing plan. Given the complex nature of managing many ships of different lengths, widths, draughts and cargoes, the operations team consists of highly trained data analysts who compare each ship against their anticipated needs, tidal data, and scheduling requirements.

Every port has facilities that are either in-house or external to the port. For instance, some ports have container vessel berths within the perimeter of the main port, with additional facilities for oil tankers located a short distance away. The Port of Southampton in the south of England is a perfect example. Alternatively, passenger berths may be located again within the main port itself, or within a separate complex altogether. Whichever way the port is designed, each facility will have its own unique and separate berthing plan. Since every port uses the same waterway, inefficient planning between facilities may create unwanted traffic and bottlenecks within the port. Therefore, it is critical for the operations team to consider every aspect of the port's operations when developing the berthing plan. Central to the berthing plan is the port's resources. These include any machinery, equipment or structures that are directly involved in the loading, unloading and ancillary operations of the ships alongside.[2] The most obvious equipment used in discharging cargo operations are the cranes (spreader and hook types), berths, quayside structures, warehouse and storage facilities, refuelling stations, etc. When designing the berthing plan resource allocation is vitally important as it charts out the choreography of movements of each facility and each vessel in attendance. For the efficient allocation of resources, it is essential to account for the size of the vessel, the number of cranes required, the size of the berth, the ship's unladen and laden draught, fuel and shoreside power requirements, and the number of tugboats needed for support.

Objectives of the berthing plan

There are four primary objectives that must be met when designing the berthing plan: (1) minimisation of vessel service times; (2) optimisation of arrival and departure times; (3) minimisation of early or delayed arrivals and departures; and (4) optimisation of fuel, berthing, and emission costs. Vessel service times refers to the duration a vessel spends in port *excluding* arrival and departure time. This includes any time spent waiting due to pre-existing traffic in the port and any vessel handling time whilst the ship is alongside. As the vessel cannot be serviced during this time, it is considered idle or lost time, where the port cannot efficiently handle the ship. To maximise revenue, it is critical that vessels are brought alongside, handled and discharged in as little time as possible. Managing vessel arrival times is another important objective of the berthing plan. Vessels cannot be scheduled to dock without the adequate facilities available. In other words, a dock without a spreader crane is no use to a vessel that requires one. The docking schedule should be staggered so that ships are able to minimise their waiting times. By doing so, port resources are maximised, waiting times are reduced, and other efficiency increases.

The arrival time plays a central role in determining servicing times and ensures that no bottlenecks are formed. To avoid bottlenecks developing, the berthing plan should ensure ships are evenly distributed between the available berths, notwithstanding, of course, any of the limitations discussed above. Another important objective is the minimisation of delayed ship arrivals and departures. These types of events have a nasty habit of throwing the entire system into disarray. For instance, if a ship arrives earlier than expected, it will have a large idling and waiting time that reduces its efficiency. On the other hand, if a ship arrives late, it will also depart late, which creates a propagating delay that cascades down to other ships. Lastly, it is important to ensure that whilst port resources are efficiently allocated, the ship must also be efficiently serviced. It is prohibitively expensive to keep a ship moored alongside longer than necessary; berthing fees, fuel, shoreside power generation and crew costs amongst others quickly add up; when a ship is lying idle, the shipping company is not making money. A port that keeps ships alongside without good reason will quickly find that shipping companies will take their custom elsewhere.

Not directly attributable to the port themselves, yet another cost shipping companies must increasingly absorb, is the *green levy*. This is an emissions charge levied on shipping lines to reduce maritime pollution. The green levy is calculated on the time a vessel spends within the territorial waters of a coastal state. The charge is designed to compensate for any pollution emitted by the vessel. despite the levy being relatively small, the longer a vessel spends within the levy zone, the higher the charge. With shipping lines already operating on razor thin margins, any increase in operational costs will likely have a negative impact on shipping company profits. Thus, the overall aim of the berthing plan is to create a timetable of each ship's arrival and departure, combined with a list of facilities and services that will need to be provided to each vessel. A well-thought-out and prepared berthing plan benefits not only the port, but also shipping companies.

Factors that influence berthing plans

In addition to the aim and objectives of the berthing plan, there are certain factors that determine the resource allocation and efficiency of the plan whilst it is being drawn up. These factors include each ship's: (1) *Expected Time of Arrival* (ETA); (2) *Expected Time of Departure* (ETD); (3) the estimated berth and operational duration; (4) the estimated crane moves per ship per berth; and (5) any miscellaneous factors, including the weather, tides, surrounding vessel activities and so forth. The ETA of a ship depends on when it departs from its homeport, the weather on route, and any traffic in the region. For the port, the ETA of each ship arriving alongside must be gradual and staggered. This ensures there is a small window of time in which each vessel can be properly serviced. The same principle applies to the ETD. If vessels arrive and depart at different times, it may become easier and quicker to manage the port's resources more effectively. If mobile machinery is required for one ship, and another ship with a similar requirement is due to arrive around the same time, it makes sense to berth both ships in the same area wherever possible. This further impacts on berth servicing times. If the ship arrives during peak hours, fewer cranes and spreader trucks will be available to handle cargo. This means it usually takes considerably longer to load and discharge vessels during busier periods. Larger ships will generally require more time to load and unload than smaller capacity ships. Servicing multiple smaller ships at the same time will free up valuable resources for larger, slower ships. On the other hand, arriving at off-peak or unsociable hours, or using smaller ports, can often allow more of the port's resources to be focused on servicing a single vessel. Although unpopular for the crew, this reduces the overall turnaround time for the vessel.

An important factor that determines the berthing plan for each type of ship is the number of crane-moves needed for unloading and loading cargo. Minimising these moves is important, as they can be considerably time-consuming. To aid and reduce crane-moves, crane operators are provided with precise details on the exact movements to be made. This means cargo operations can be completed faster. Last of all, there are various miscellaneous factors that can affect the speed at which a vessel is serviced when in port. These include weather patterns, tide schedules, and the volume of local marine traffic. All these factors must be accounted for when preparing the berthing plan.

Sample berthing plan and detailed explanation

Generally, berthing plans are tabular in nature. The columns indicate the days and their respective hour slots. The rows are subdivided into two categories. The first category is based on the number of berths available. So, each berth will have its own row with specific berthing details. The second category is contingent on each berth and contains data specific to the vessel requirements for that berth. Among the sample requirements are the number of cranes needed, the number of containers to be loaded or unloaded, and any additional notes relevant for the cargo superintendent and port operators. Below (Table 13.1) we have a sample berthing plan. The first observations we can draw from this plan is that the rows and columns display a detailed report on the ship and its berthing window. We can see that a vessel (*Torrens*) is occupying berth 1 from between 0000–0600 to 1800–2400 on day 1. On day 2, berth 1 will be similarly occupied by vessel 2 (*Ocean Carrier*). For a certain duration. Alternatively, we can deduce that vessel 3 (*Morning Glory*) will be occupying berth 2 for the entire two days. In general, the time column is divided into six-hour slots, i.e., the day is divided into four equal parts. This makes managing the plan a lot easier. The row with the entitled NAME holds the vessel name and its Flag or country of registration, though the latter part is not always required. The row below the ship's name and Flag is the ship's callsign. This is the unique identifier that each individual vessel must use for radio transmissions. Below that are the number of cranes required. This part is especially important. If we take the example of *Torrens*, we know two cranes are needed between 0600–1200 and 1200–1800. The number two signifies two cranes are needed fore and two cranes are needed aft. Ships often have different crane requirements depending on the size and number of containers to be loaded and unloaded. In the bottom row we can

Table 13.1 GMDSS Areas, Ranges and Equipment

#	Time Slots	Day 1				Day 2			
		0000–0600	0600–1200	1200–1800	1800–2400	0000–0600	0600–1200	1200–1800	1800–2400
BERTH 1	NAME	TORRENS (NO)				OCEAN CARRIER (UK)			
	CALLSIGN						UNITED KINGDOM		
	CRANES		2	2			3	2	
	CNT. FLUX		–130 /+50				–50 / +140		
BERTH 2	NAME	MORNING GLORY (JP)							
	CALLSIGN								
	CRANES			3	2				
	CNT. FLUX		–200 / +30						

see how many containers must be unloaded, denoted as positive, and the number of containers to be loaded, denoted as negative. Although this convention may vary from port to port, it provides an indication of the flux of containers within any given port.

As more containers are unloaded, there is a positive influx of containers into the port. As containers are loaded, there is an outflux of containers from the port. Using this information, the port operators can manage where and when each container is stowed at any given time.

The digitisation and automation of berthing plans

The process of drawing up the berthing plan used to be done manually up to a few years ago and, as might be expected, was a challenging and often tedious undertaking. With the development of specialised software port operations can be mathematically modelled to represent the influx and outflux of ships and containers. This has made the job of port planners much easier and vastly improved the efficiency of port operations. That said there are always various ways to compute the ship-to-berth combination, meaning computations must be performed several times before the most effective allocation of resources can be found. For this reason, ships had to inform each port of arrival at least six months in advance about any planned stops at port. The process was tedious and consumed several man-hours spent on administration and paperwork. For liner vessels this is not such a problem as the ship visits the same ports along a set schedule. Calculating estimated dates and times of arrival and departure were relatively straightforward. For non-scheduled ships, however, the situation is completely different. These ships go from one port of call to another collecting whatever cargo is booked at a specific date and time. These ships operate on the *spot market*; they don't follow a specific schedule and will literally go anywhere the customer demands. As we can imagine, it would be almost impossible for such a vessel to plan a week never mind six months in advance. Today, however, digitisation has made this task much easier meaning reports can be sent electronically in real time.

The system automatically accounts for new ships added to the incoming database and provides a suitable berthing plan at the touch of a button. The software works on a repeated brute force algorithm that finds the optimal allocation while following complex mathematical models to reduce computational time. A major advantage of using such modelling software is that they can detect patterns in berthing plans and ship schedules. Since the global shipping fleet can be classified into a few broad types of vessels that require the same resource allocations, the system is able to assign a unique identification for each type. That way, if there are two tankers of similar sizes, they are registered as vessels that require the same allocation of resources. Over time, the port berthing plan and schedule of incoming vessels will begin to repeat. While it is difficult for humans to access records of the past several years and recognise these patterns, computational software can perform this task in a fraction of the time. The software analyses existing berthing schedules for past patterns, and then applies a tried and tested model. Thus, instead of having to create a new berthing plan every time, a quick historical pattern search will yield an optimised solution based on past performance.

So far in this chapter we have seen how the berthing plan is a critical document used by ports to allocate resources efficiently and effectively. The role of the mariner in this task is necessarily limited to advising the port authorities of the date and time the vessel is due to arrive, and the number and type of containers or cargo to be loaded and discharged. There is, of course, a whole host of other information vessels must provide before arriving in port but suffice it to say as far as cargo management is concerned, the berthing plan is the critical document. In the next part of this chapter, we will turn our attention back to the duties and functions of onboard personnel involved in bringing the vessel safely into port.

Mooring operations

The mooring operation is one of the important tasks that seafarers must perform on ship's deck. Technically, the operation may seem simple but there are many dangers and hazards associated with mooring. Working on the ship's deck is not an easy task. Officers and crew must follow a host of health and safety regulations and practices that are designed to minimise the risk of personal injury when operating deck machinery. When undertaking mooring operations, however, the level of danger increases exponentially with the entire working area becoming a potential *death trap*. This means that additional safety considerations must be followed. First and foremost is prohibiting unnecessary crew members on deck during the mooring operation. Only those members of the crew who are directly involved in mooring the vessel should be at the mooring station. Anyone not involved in the operation should be removed to a safe distance, or preferably off deck entirely. The second main consideration is the prevailing weather conditions. Before planning a mooring operation, always consider the weather and wind. Wet slippery decks and ropes will make handling that much harder and potentially lead to trips, slips, and falls. The wind and currents will also adversely affect vessel handling. The master and responsible officer must ensure they check the current and future weather conditions before authorising the crew to undertake mooring operations.

The third consideration concerns *rope bight* and the *snap back zone*. All personnel involved in the mooring operation should be acutely aware of the hazards associated with rope bight and snap back zones. Mooring ropes are long and heavy and stored onboard in a coil. When these ropes are placed under stress, they tend to form a coil or ring shape called the *rope bight*. If a person involved in the mooring operation falls within the rope bight, the pull of the rope can drag the person overboard or smash them into the on-deck machinery. Alternatively, the highest number of injuries and crew member deaths suffered during mooring operations concern the *snap back zone*. Mooring ropes and cables are made using several individual strands which are then wrapped around each other to form a strong and durable rope. As these ropes age, they begin to weaken. If a rope is under stress and it suddenly fails, it will whip back at great speed and in an uncontrolled manner. Anyone unfortunate enough to be standing in the way of the snapped rope will likely sustain life-changing injuries or worse. The area travelled by the rope is referred to as the snap back zone. It is important to recognise that when ropes are pulled straight then the snap back zone is minimal, whereas if the ropes are turned in to a bollard or roller then the snap back area will increase. The marking of snap back zones on ships, although convenient and simple, does not reflect the actual complex trajectory a snapped rope will travel. In the past this has led to many seafarers feeling they are safe if they are not in a painted area. Of course, this is completely erroneous. To combat this, the 2015 version of the *Code of Safe Working Practices for Merchant Seafarers* (COSWP) revised its guidance on snap back zones and now dissuades the marking of snap back zones on deck. Instead of marking out specific hazard areas, the revised guidance recommends treating the entire mooring deck as a potential snap back zone, with appropriate signage to warn crew members they are entering a hazardous area.

Fourth, is checking the mooring equipment for faults and defects. This includes the mooring winches, load sensors, drums, windlass, ropes, and cables. Although these should be checked regularly as part of a preventative maintenance regime, it is not uncommon for even obvious defects to be missed during standard maintenance inspections. Some mooring lines are fitted with a *tail*. This is a short length of synthetic fibre which is placed in series with the vessel's winch mounted wires to decrease mooring line stiffness. This in turn reduces the peak line loads and fatigue caused by the ship's motions. When checking over the mooring equipment, it is important to inspect the tails for each service (breast, spring, and headline) and to ensure they are all the same size and material as different tail lengths and materials may lead to an uneven load on the mooring line. The fifth point to note is only tend to one line at a time. It is of paramount safety that only one line is tended to at any time during the mooring operation. If not, increased loads may develop on the other tended lines. If two lines are tended together it could lead to an overload and snap back. The sixth point is maintaining a visual check on the mooring line load. Ensure that the allowable breaking load on any of the mooring lines does not increase by 55% of its Maximum Breaking Load (MBL). This is done to prevent the line from breaking. Seventh, always avoid mixed mooring. Mixed mooring is extremely dangerous and is the result of using lines of different sizes and materials. Ideally, all mooring lines should be the same length, width and made from the same material. If this is not possible, then all lines of the same service must be the same – for instance, all breast lines must be the same or all spring lines must be the same. The use of mixed moorings comprising full-length synthetic ropes used in conjunction with wire must be avoided wherever possible. If a synthetic rope and a wire are used in the same service, it is the wire that will carry the entire load while the synthetic rope will carry practically none putting the entire strain on the wire alone.

The eighth point to note when mooring is to arrange the mooring lines in a symmetrical pattern. All mooring lines should be arranged as symmetrically as possible with the breast line. The breast line should be perpendicular to the longitudinal centre line of the ship and the spring line should be parallel to the longitudinal centre line. This means the weight and strain of the vessel are shared and any vessel motions are dissipated equally through each of the moorings. Last, but certainly not least, is the final and nineth point: keep a continuous check. The loading on the mooring lines must be checked continuously even after the mooring operation is complete. Whenever there is any change in the ship's ballast condition, the lines must be slacked or tightened accordingly. Likewise, when there is a change in draught caused by rising or ebbing tides, the mooring lines must be adjusted to account for the rise or fall in the vessel against the quayside. Granted, this will not affect larger vessels, but smaller boats and watercraft may find themselves hanging precariously if tidal changes are not accounted for.

Anchoring

Anchoring is an important operation that ships perform to prevent them for drifting on ocean currents. There are several situations where a ship may need to drop anchor. When doing so, it is critical the operation is done efficiently and in accordance with the correct procedures. The anchor is a piece of marine equipment that is designed to restrict vehicle or structural movement in water. The anchor achieves this by either using its weight to hold the structure in place, by clamping down into the soft bed of the waterbody, or a combination of the two. In addition, anchors can also act as a *drogue* or a positive drag mechanism for ships in stormy weather. The drogue helps to keep the vessel stable and steady and prevents the slamming of the bow or green water flooding caused by temporary bow and deck immersion.[3] Although naval architects generally try to reduce drag as much as possible when designing ships to increase their straight-line speeds, drag can slow vessels down during heavy seas so that they remain under the control of the propulsion system. This prevents the rolling motion of waves from damaging the vessel. Traditionally, anchors were found on board large ships such as container vessel and oil tankers to hold them in place while being moored or if the vessel was required to come to a standstill mid-sea. Today, many other ocean structures deploy anchors for the purpose of staying stationary; for example, oil rigs, offshore platforms, and floating production, storage and offloading (FPSO) units also make extensive use of anchors. Unlike ship anchors, though, these tend to be semi- or permanent anchors designed to keep the structure in place for extended periods. Ship anchors, by comparison, are necessarily temporary.

Anchors are generally made of metals that are made resistant to long-term corrosion through electroplating and galvanisation. Some anchors may be manufactured from fibre-reinforced composites or polymers (such as carbon fibre). The advantage of using such materials is their high strength-to-weight ratio. This means that compared to traditional metal anchors, even light reinforced anchors can withstand enormous levels of load and strain. The main disadvantage though is the high cost of development and production. Furthermore, for heavy structures such as ships, such lightweight anchors will provide a negligible effect for holding the structure still, especially in heavy swells or high sea states. Even so, anchor technology continues to develop unabated with new research investigating the use of multilayered crossed fibres in composites to provide improved weight properties without adversely affecting the integral strength of the anchor.

Most ships over a certain size[4] use temporary anchors to keep them in place for a short period of time. Generally, they are used to berth and moor vessels or to stop them dead in the water. Unlike permanent anchors, simply attaching a deadweight to the anchor line is not practical as reeling the anchor back in then becomes impossible. Subsequently, temporary anchors rely on the clamping or hooking of the anchor onto the seafloor or ocean bed. To achieve this, the anchor uses a combination of its own weight and gravity to drive and embed a set of *flukes* or pointed shafts into the soft bottom of the waterbody. The majority modern temporary anchors derive their design from the fluked, admiralty and stockless anchors. Since temporary anchors are used to tether vessels for a short period of time, there is no requirement to manufacture them from expensive materials such as titanium. As the anchor can be hauled back onto the vessel in a matter of minutes, they are often made from metals and alloys that do not easily rust and are coated in a layer of zinc or some other anti-corrosion material. To remove the anchor from the seafloor – or to dislodge it if caught in rocks or underwater structures – tripwires are used to provide an additional moment of pull. These also serve as an additional source of force that can haul the anchor head quickly to the surface should the need arise. To embed the anchor, it is let down slowly till it reaches the seafloor. At this point the ship is still slowing down. This is important as for the anchor to work properly, it must catch onto subsurface debris, rocks, or cracks in the seabed before the vessel comes to a complete stop. Once the anchor has gained a hold, it must be permitted sufficient time to settle in to prevent motion caused by light waves and subsurface currents. In most cases, for ships that have set down their anchor in open sea, small movements caused by waves, wind and currents are largely inconsequential. However, for moored ships or vessels berthed alongside, they must not shift from their position. Any movement of the ship can cause damage to the hull or the quayside. To prevent this from happening, tugboats and additional mooring lines must be used to support the anchor in holding the vessel steady.

Figure 13.1 Ship weighing port side anchor.

Mooring buoys

Mooring buoys are a type of buoy to which ships can be temporarily moored. They typically weigh more than standard navigational and marker buoys and have a heavy weight located beneath which anchors the buoy in position. Attached to the buoy are loops or chains to which the ship can attach itself and remain stationary without employing the ship's anchor. Mooring buoys are often used in protected or fragile marine environments such as coral reefs. Because anchors are destructive to the seafloor, by using mooring buoys, ships can be held in position without upsetting the local marine ecosystem. Mooring buoys are predominantly found in sea areas with a rich marine environment such as off the coast of Australia, throughout Asia and in some parts of the United States. Alternatively, mooring buoys are also often found around ports and oil tanker-discharging centres. Due to the inherently dangerous nature of their cargo, most port authorities are understandably reticent about allowing oil tankers within their port confines. To get around this, tankers are often tethered to special types of mooring buoys which are positioned some distance offshore. From here, the tanker can either disgorge its cargo directly into pipes which pumps the oil onshore, or to a second smaller tanker which positions alongside. This process is called *ship-to-ship transfer*.

Figure 13.2 Fo'c'sle detail of a container ship passing through the Panama Canal.

Types of mooring

Every vessel is designed and built with mooring arrangements. These are designed so that a vessel can be positioned alongside a berth, between two or more mooring buoys, to a single mooring buoy, to another ship or barge or against a jetty. Each of these mooring operations requires very different mooring techniques. In essence, there are 10 primary methods of mooring:

(1) Mediterranean
(2) Baltic
(3) Running
(4) Conventional
(5) Standing
(6) Spider
(7) Single Point
(8) Single Buoy
(9) Multi–Buoy; and
(10) Ship to Ship.

Mediterranean mooring

What happens when there is insufficient space in port for the number of ships to be moored? Or, what if there is a ramp at the stern of the vessel which is connected to the ship's hull (such as with car carriers)? These situations present a unique challenge for port operators and the ship's officers. The answer is to moor the vessel perpendicular to the quayside with its stern transom (surface) lying parallel to the quay. This process is called *Mediterranean mooring*. Mediterranean mooring, or 'med mooring', is a technique for mooring a vessel to a pier at a perpendicular angle. This means the ship occupies less space as it is connected to the pier at its width, rather than its length. The disadvantage of this technique is that collisions are more likely to occur, and it is impractical in deep water or in regions with large tide differentials. Mediterranean mooring can be achieved in two ways: *berthing stern to*, or *berthing bow to*. When berthing *stern to*, the ship is reversed into the mooring position. For vessels with a standard right-handed propeller, the dock is best kept to the port side. The ship is first kept parallel to the berth and is gradually brought astern. The point of berthing is now generally abeam. Throughout the astern movement, the rudder is kept hard over port as to generate maximum canting effect. As the ship falls astern, the bow will naturally cant to starboard due to transverse thrust. To counteract this, the starboard anchor is dropped in such a way that the scope (the ratio of the length of the anchor rode[5] and the vertical distance from the fo'c'sle to the seabed) is approximately 8:1. This means if the depth of the water is 10 metres (32 ft) down from the bow, a total rode distance of 80 metres (262 ft) is needed. The chain is left to free fall to allow the anchor to dig into the seabed. It is important to ensure there is no snag in the rode and that there is sufficient length of the chain to reach the berth. Equally as important is to ensure the anchor is dropped away from any other anchor lines or lazy lines (which are used as a substitute to the anchor) otherwise fouling of the anchor chains may occur.

Whilst the vessel is making sternway, if not already in place, hand fenders should be arranged along the side of the vessel to protect the hull from colliding into other vessels. The vessel is slowly brought astern. Offshore winds are helpful in dampening the stern board movement of the vessel, which reduces the need for ahead manipulation of the telegraph. When the wind is athwart, the first line ashore is from the windward side. In this case if the wind is pushing the vessel to starboard the starboard quarter line must be sent ashore first followed by the port side quarter line. Both lines must then be *doubled up*, i.e., the lines are made to take a turn on the shore bollard and sent back onboard. This helps the mooring crew whilst hauling out as only one end of the casting line need be retrieved by the winch. No shoreside assistance is therefore required. During windy conditions, it may be necessary for the first shoreline to be sent ashore by a small motorboat. This means the vessel can be warped in with the help of its winches. Whilst hauling in, if the bow swings, corrective action must be taken by way of the bow thrusters. Once the vessel is berthed in place, the lines ashore are made fast in such a manner that the angle between them is sufficiently wide to restrict any stern swing caused by wave or wind surges. It is always advised to set an additional pair of lines across the stern to shore for added security. Now the boarding ramp can be lowered, and cargo loaded or discharged accordingly. Docking *bow to* is comparatively easier to berthing *stern to* as the master and pilot has much more control over the vessel's movements. In this scenario the vessel approaches the berth bow first. A line from each quarter is hitched to the chain aft to restrict any sideways play. Two lines are passed ashore from the port and starboard bow at wide angles and made fast.

Baltic mooring

Baltic mooring is a mooring technique that can be used when a ship is berthing without the assistance of tugboats during heavy winds. In this situation, the master uses the ship's anchor in addition to the mooring lines to keep the vessel securely moored. The vessel is berthed alongside the quay by way of the stern line, which is shackled to an offshore anchor cable in the region of the *ganger length*.[6] When approaching the berth, the offshore anchor is deployed with the weight of the cable and the stern mooring acting in unions to hold the vessel just off the quay. To perform the Baltic mooring technique the following procedures must be followed:

(1) First a 30 mm wire is passed out from the poop deck on the offshore side from the outer side of the hull and clear of any protrusions such as the gangway, pilot ladder, and so on

(2) The anchor is then cockbilled, i.e., released a little way from the hawsepipe before being let go

(3) A member of the crew is lowered using the bosun's chair[7] to tie up the wire to the anchor with a shackle at about the ganger's length

(4) The opposite end of the wire is taken *on turn* upon a mooring winch through a bight

(5) When the ship is abreast of the berth and falling on it rapidly, the anchor is dropped keeping a trickle of headway so that the anchor holds

(6) Once the anchor has snubbed, the wire from the stern that goes in with the anchor is made taught. This effectively holds the fall of the stern

(7) The anchor chain is then slowly paid out simultaneously with the holding wire. At the same time, the onshore wind will push the vessel horizontal to the berth

(8) As soon as the vessel is close, the spring, head and stern lines are passed ashore with the heaving lines. The scope of the anchor is adjusted accordingly to bring the ship slowly alongside the berth. Typically, the anchor is dropped some 21 to 30 metres (70 to 100 ft) off the berth depending on the force of the wind and the tonnage of the vessel.

Single point mooring

Single point mooring (SPM) is a type of floating buoy or jetty which is anchored offshore to allow the safe discharging of hazardous cargoes such as crude oil. SPM is primarily used in areas where a dedicated facility for loading and unloading liquid cargoes are not available. They are usually located some distance from the shore

Figure 13.3 Single Point Mooring Spm Sokol.

facilities and are connected by way of subsea pumps and pipelines. The main benefit of SPM is that even colossal-sized tankers can be accommodated where shore space is limited, or marine traffic is high. The other benefits to SPM include ships with high draughts can be moored easily and without risk of running aground, large volumes of liquid cargo can be handled quickly and safely, and there is no requirement for the ship to enter port which saves considerable time, effort, and money.

As stated above, the SPM is a special type of offshore anchored mooring buoy and is divided into different sections. Each section serves a specific function. The main components of an SPM are the mooring and anchoring system, the body of the buoy, and the product transfer system. The SPM is moored to the seabed using a mooring arrangement that typically includes one or more anchors (either gravity based or piled), anchor chains (*legs*), and chain stoppers. The chain stoppers are used to connect the chains to the buoy. This mooring arrangement is designed to allow the buoy to move freely within defined limits. This prevents the buoy from potentially breaking free during high winds, heavy sea, strong currents, and, of course, the movement and sway of the tanker. The part of the SPM which floats above the water's surface has a rotating part which connects to the tanker. This rotating part allows the tanker to stabilise its position around the buoy. The tanker is usually moored to the buoy by way of a hawser arrangement, which consists of a nylon or

Figure 13.4 Single Point Mooring Spm Sokol.

polyester ropes shackled to an integrated hook on the buoy deck. Chafe chains are connected to the tanker end of the hawser to prevent damage from the tanker fairlead. This process follows the procedures established by the OCIMF.[8] The product transfer system is located at the heart of the mooring buoy. The system transfers the liquid product to the tanker via a Pipeline End and Manifold (PLEM) located on the seabed. Flexible hoses, called risers, connect the subsea pipeline to the buoys product transfer system. The buoy is connected to the tanker using floating hose strings, which are fitted with breakaway couplings. These are specially designed couplings that disconnect at a predetermined break load; when the break load is reached, a series of internal valves activate which automatically shut off both ends of the hose, preventing any spillage of product. The SMP system uses a swivel system which connects the PLEM to the buoy. The product swivel system provides flexibility and freedom of movement to the tanker during the transfer of product. This movable pipeline connection prevents premature hose failure caused by traction or bending stresses.

As a general overview of how the SPM system works, the tanker is moored to the buoy for loading or discharging of product. A boat landing space on the buoy provides access to the buoy for setting up the relevant connections and for securing the ship. Fenders help to protect the buoy from the movements of the ship, especially in inclement weather. A set of lifting and handling equipment on the buoy facilitates the positioning of the hose connections. Once the connections are made, valves are operated remotely from an electrical substation. This ensures the necessary safety alarms and navigational aids remain fully operational throughout the loading and unloading procedure. The liquid cargo is transferred from the geostatic location (i.e., the PLEM) to the tanker using the buoys SPM system. Once the product is transferred, the tanker is disconnected from the buoy. All systems are returned to their static condition and the tanker is free to continue underway.

Multiple point mooring

With this method of mooring, the bow of the ship is secured using both anchors and the stern is secured to a buoy. When readying to berth, the vessel approaches the berthing position from ahead at an angle of 90 degrees to the final direction of mooring. The starboard anchor is let go first at a predefined location whilst the ship is making headway. A calculated length of cable is paid out, and the telegraph is set to astern to slow the vessel to a complete halt. Once the vessel is stopped, the port anchor is let go. This allows the vessel to position her stern along the centre line, bifurcating the buoys. For aligning the vessel along the centre line, the port cable is paid out and the starboard cable is heaved in with astern propulsion. The helm and engines must be carefully manipulated during the manoeuvre to ensure the stern swings clear of any obstructions including the buoy. During unberthing, the anchor cables are heaved in to move the vessel forward. The weight is taken on the windward lines whilst casting off the other lines to prevent the stern from swinging. This manoeuvre requires skill and experience to perfect it into a fine art. The weight on the lines can be immense and so there is a real danger of the lines snapping or the stern of the vessel colliding with the buoy.

Conventional buoy mooring

The multiple point mooring procedure can be used with more than one buoy which are permanently fixed to the seabed. This is referred to as the *conventional buoy mooring* (CBM) technique. In these situations, there may be between three and six buoys that are permanently held in place in a rectangular pattern. This series of buoys act as a strong point to attach the ship's mooring lines. The multiple buoys or CBM system is useful when there is no quay available, or the ship is unable to manoeuvre into place. it can also be used in conjunction with a fluid transfer system which enables the connection of subsea pipelines to the midship manifold on standard tankers. In the case of the latter, the submersible hose or hoses are stored on the seabed. For cryogenic liquids,[9] the hose is suspended from a tower to the midship manifold. Whereas some buoys are 'off the shelf', others are specially designed to include features such as quick disconnection couplings and plugs. The mooring system and layout of the buoy are always specific to match the type of vessel that will use them, as well as considering environmental and marine conditions. Typically, but not always, mooring buoys are found within inshore environments where the water depth starts from about six metres (19 ft) from the water's surface to the seabed.

Ship to Ship Mooring

Ship to Ship Mooring (SSM) is distinctly different from the other types of mooring discussed in this chapter as it involves two or more ships mooring alongside each other. SSM is predominantly done during cargo transfers, when either one vessel is anchored or both vessels are underway. The mooring arrangement depends on the size of the ships involved. For perform SSM, a vessel either at anchor or stopped and maintaining a constant heading is approached by the manoeuvring ship. The latter must approach the stationary vessel at as small an angle as practicably possible. The approach is usually done abaft the beam of the constant heading. During the approach, as the manoeuvring ship closes in, it steers a course parallel to the heading or course of the other ship, reducing the horizontal distance between the two ships to less than 100 metres (328 ft). Once the manoeuvring ship is in position, she must use the engine and rudder to reduce the distance further until the fenders of both vessels are touching. After the two ships have made parallel contact, the lines are exchanged as per the mooring plan. It is common practice during the approach for the wind and current to be ahead or at least at very small angles to the bow.

Running and standing mooring

Vessels must sometimes use the seaward anchor in conjunction with the mooring lines to haul the vessel away from the berth when casting off. The seaward anchor may also be used to control the rate of lateral movement towards the berth when the vessel is coming alongside. Both these manoeuvres can be made with or without the assistance of tugboats. There are two situations where this process may be applied: *running mooring* and *standing mooring*. Compared to Mediterranean mooring, running mooring takes a relatively shorter time to complete and offers more control over the vessel's movements. The ship's starboard anchor is let go at a position approximately four to five shackles from the final position of the bow with around nine shackles paid out whilst the vessel is moving ahead under engine power. As the ship falls astern with the tide, the port anchor is let go and the starboard anchor is heaved onto five shackles. This method restricts the ship's swinging room and reduces the load on the windlass. Alternatively, *standing mooring* is practiced during cross wind conditions. As the vessel is stopped, the port anchor is let go and with the tide, around nine shackles are paid out. The starboard anchor is let go at the same time as the port anchor is heaved in. Thus, the port anchor is held at four shackles – generally being the flood anchor – and the starboard anchor is held at five shackles – as the ebb anchor. This operation takes considerably longer to complete and provides far less control over the vessel's movements. It also places substantially more load and stress on the windlass when compared to the running moor procedure.

Tugboat operations

A tugboat or tug is a type of secondary boat that assists in the mooring and berthing of a ship by either towing or pushing the vessel towards the dock. A tugboat is a special class of boat without which it would be nigh on impossible for today's megaships to make it into port. Together with their primary role of towing the vessel towards their berth, tugboats can also provide auxiliary services such as providing water, air, and supplies. When manoeuvring a vessel into port, the tugboat eases the client vessel by pulling or pushing the vessel along. The tugboat helps keep the larger vessel on a straight course; this is particularly important when navigating through narrow or congested waterways. The unique component of the tugboat is their propulsion system. This is specially designed to provide the tugboat with the power and torque to move larger and heavier vessels whilst retaining a comparatively smaller size. This makes the tugboat especially nimble. The use and function of tugboats varies from one port to another, and different port operators will have different requirements according to the type of traffic they receive. The average tugboat has an engine that can produce between 680 and 3,400 bhp (roughly equivalent to 500–2,500 kW). Deepsea tugboats, however, can produce as much as 27,100 bhp (20,000 kW) with a power to tonnage ratio ranging between 2.20–4.50 for larger tugboats and 4.0–9.5 for harbour tugboats. These extremely high ratios far outmatch the power to tonnage ratio of the average container ship, which ranges between 0.35 and 1.20. To provide the power tugboats need, they have engines that are quite like those used by railway locomotives. The main difference being the tugboat engine drives a propeller instead of producing an electric motor supply.

In line with their awesome pulling and pushing power, tugboats are generally classified according to one of two categories:

(1) Escort tugs, which are designed to escort ships into and out of port and manoeuvre them alongside; and
(2) Support tugs, which provide support and towing services to the offshore industry.

Irrespective of their category, tugboats are further divided into three types:

(1) Conventional tugboats
(2) Tractor tugboats; and
(3) Azimuth stern drive tugboats.

Figure 13.5 Container Ship *MSC Charleston* leaving the Port of Bremerhaven.

Conventional tugs

Conventional tugs are fitted with diesel engines and either one or more propellers. Single-propeller tugs are further categorised into two classes: right-handed conventional tugs and left-hand conventional tugs. The right-hand conventional tug is the more common class and is used extensively throughout the world for their reliability and high-power efficiency. The main components of the conventional tug include:

(1) A classic rudder arrangement
(2) The towing hook, which is usually positioned in the centre of the tug;[10] and
(3) A stern mounted powerplant complex.

Central to the tugboats' operational capability is the configuration of the propeller. On conventional tugboats, this is always a screw propeller. This propelling device is diesel-driven, and as it spins round, it creates thrust. This thrust is what moves the boat forward or aft. Tugboats are designated by the number of propellers they carry. For example, a single screw tugboat means it has one propeller, whereas a twin-screw tugboat would have two propellers and a triple-screw tugboat would have three propellers, and so on. Whilst the tugboat may be fitted with moveable nozzles and rudders, the propeller arrangement itself is immobile and fixed. There are various key advantages to the design and features of conventional tugboats. First, the design and construction of conventional tugs is straightforward, which means they require less maintenance. This makes them cheaper to operate and maintain compared to the other types of tugs available. Because conventional tugboats are self-sufficient, they require no additional support systems. Lastly, their extreme power and manoeuvrability makes them idea for operating in congested waterways and tight harbour basins. Like all things, though, where there are positives, there are also disadvantages. The main issue with conventional tugboats is that they are unidirectional. This means they cannot work in a reverse direction, unless fitted with a reduction gear to facilitate reverse flow. Conventional tugs are known to suffer from an increased risk of cavitation. This adversely affects the stability and structural integrity of the boat. Because conventional tugboats tend to follow older design principles, they lack the technology and ability to assist mega ships. This means conventional tugs are limited to supporting small and medium-sized vessels. Finally, there is also an increased risk of capsizing. To reduce this risk and to prevent the tug from girting,[11] an additional release hook is required.

Tractor tugs

The key component of the tractor tug are the two multidirectional propulsion units, which may consist of large rotating outboard motors or rotating vertical blades. In either case, the thrust units of the tug are placed side by side and under the bridge. This means the towing point can be placed much closer to the stern, which maximises the available propulsion power. The benefit of this configuration is that the thrust is always outside the towing point, thus creating positive turning moment. A rotating disc decides the magnitude of the force of thrust. The versatility of the tractor tug is further enhanced by working the tugboat's towline directly from a winch drum with a remote-control joystick on the bridge. The tug master can thus alter the span of the towline according to the needs of the task. With their precision manoeuvrability, tractor tugs are by far the most widely used class of tugboats. With them being the most popular, tractor tugs have several advantages and benefits over the conventional class. The most important benefits are:

(1) The ability to provide full thrust over 360 degrees
(2) A rapid power-on response time and excellent manoeuvrability
(3) In direct contrast to the conventional tug, the tractor tug is far more adaptable and can respond to changes in demand easily
(4) The control systems are very simple and pose little risk of girting or capsizing
(5) Because of their power rating, tractor tugs offer extraqrdinary performance and speed

(6) Tractor tugs can overcome the interaction forces when coming into close quarters with larger vessels

(7) Tractor tugs are able to work efficiently in a sideways movement in part to the close proximity of the propulsion unit to the turning point and in part to the absence of the rudder. Despite their benefits, tractor tugs do have some negative attributes as well.

These include:

(1) Having less bollard pull than Azimuth Stern Drive (ASD) tugs (see below)

(2) Tractor tugs are extremely expensive to purchase and maintain, meaning they require high capital investment

(3) The complex subsea arrangement is costly to maintain and repair

(4) The short distance between the pivot point and the thrust creates a short turning lever, meaning tractor tugs are not overly suitable for open water operations

(5) Tractor tugs can generate a full side thrust of up to 21 degrees; this means there is a heightened risk of sustaining substantial damage when lying alongside another vessel

(6) The draught on some tractor tugs can be as much as five metres (16 ft), which risks causing hull damage when running aground

(7) Tractor tugs require very specific initial and ongoing training; this adds an additional cost when retraining conventional tug masters.

Figure 13.6 Tug *Michel Hamburg* assisting *Cosco Shipping Nebula*.

Azimuth Stern Drive (ASD) tugs

These tugs are a midway between conventional tugs and tractor tugs in the sense that utilise many of the same systems and technologies. Azimuth Stern Drive (ASD) tugs can have two towing positions: one forward and one amidships. Propulsion is from two rotating azimuth units which are located beneath the stern quarter but, unlike conventional propellers, spin round on a 360 degree axis. ASD tugs are considerably more efficient than conventional tugboats but less efficient than tractor tugs. Even so, they boast impressive credentials. Unlike conventional tugs, ASD tugs have multidirectional capability; they have a more suitable hull form for open waters and can work in closed waters; they have improved bollard pull; the azimuth units are easy to withdraw for maintenance and repairs; the maximum heel with side thrust is less than 15 degrees (tractor tugs can be as much as 21 degrees); and they have a shallower draught, typically being around three metres (9.84 ft). By contrast, ASD tugs are much more difficult to manage and work with compared to tractor tugs. When operating at full power, it has been known for ASD tugs to suffer severe squat at the stern and flooding of the aft deck. They are also susceptible to girting or capsizing when using the aft towing position. Finally, it is not uncommon for as much as 99% of towing power to be limited to the forward towing position only.

Other types of tugboats

In addition to the main three types of tugboats outlined above, there are a host of other types of tugs, many of which specialised functions and systems. For example, there is the tractor tug with cycloidal propellers. Prior to the Second World War, tugboats were designed with high power ratings to maximise potential tonnage capacities. These boats required a special type of propeller called the cycloidal propeller to provide the necessary manoeuvrability. Today, these have been replaced by the Voith Schneider propeller (VSP), which is a technological advancement on a cycloidal drive. It provides excellent manoeuvring and smooth handling, making it ideal for ferries and barges. The Carousel Tug was developed by the Dutch and emerged as an award-winning maritime innovation in 2006. It consists of interlocking inner and outer rings in which the former is connected to the boat, and the latter to the tug's body. The ship under tow us connected via winch or hooks. The Reverse Tractor Tug has a Z-drive aft-mounted propulsion unit. These tugs do not have a skeg and work efficiently as escort tugs. With Combi-Tugs, a bow thruster and steering nozzle is fitted to a conventional screw tug to offer improved manoeuvrability. Z-PELLER tugs have two towing positions: one forward and one amidships. Main propulsion is provided from two rotating azimuth units which are placed similarly to the conventional twin-screw tug. The azimuth propulsion unit replaces the conventional shafts and propellers, which allows 360 degree rotation about the vertical axis. The Giano Tug is a highly capable class of tug that serves well as both a support and escort tug. It is a technologically advanced tug that allows remote manoeuvring through VSAT or 4G connections. Its 360° rotation and excellent side-stepping speed places it at the top in the order of tugs standard. Hybrid technology tugs or tugs that use Liquid Natural Gas (LNG) as their running fuel are categorised as eco-tugs. These tugs serve the same purpose of escort and support as conventional tugs but are considered more less polluting to the marine environment. Last of all, ice tugs are used to escort ferries and barges through ice frozen waters. Though not classified as ice breakers, the strengthened hull means they can break through ice packs that would ordinarily cause damage to standard hulled vessels.

Berthing manoeuvres

The berthing manoeuvre is the process where a ship is brought safely alongside a berth. Before starting the procedure, it is critical to account for the effects of wind, tide, state of the ship's trim, draught and freeboard, the ship's equipment, and any aids to navigation and manoeuvring the ship may have onboard. The master also needs to account for the support and assistance provided by tugboats. For each berthing operation, the master must assess the prevailing conditions and the ship's profile. From there, the master can devise their berthing plan. The berthing plan needs to be flexible and there should always be contingencies in case any step of the plan doesn't work out as anticipated. Although the master is always in overall command of the vessel, the pilot is on board to assist the master in achieving a smooth and problem-free manoeuvre. We will discuss the role and function of the pilot and the relationship between the pilot and the master later in this book, but it is critically important that the two work together to ensure the ship is handled effectively. Amongst many other things, communication between the pilot and master is key; this typically includes sharing information regarding the minimum water depths, tidal information, current information, general conditions of the

Figure 13.7 CMA CGM *Turquoise* entering Victoria Quay, Melbourne, Australia.

berth, tug-handling instructions, the mooring arrangement for the vessel (including the length of lines and certified bollard strength), the use of anchors, thrusters and or tug boats in the event of surge or swell, and any other relevant information that may impact on the ship's handling.

Turning the ship

The master, when planning berthing manoeuvres, must take into consideration the above factors as well as the effect of wind and wind direction on the ship, relative to its trim, draught, and speed, along with factors governing the centre of turn and positions for securing tugs. All these factors influence the centre of turn of the ship, which is the pivotal point about which the ship will rotate.

Tug escort and support

When entering the confines of the port the master may be met with a host of different scenarios. Ukc, weather, wind, current, tide, berth length, the distance between the vessel's forward and aft positions at the berth, and the size and depth of the turning basin are all integral factors. Based on these, the master can determine what, if any, tug support is needed. For most vessels, this means the standard centre lead forward and aft escort. This configuration controls both the lateral and forward movements of the ship, and the aft movement by pulling or pushing as required. If only one tug is used, then this is usually secured aft. In these situations, the tug is exclusively used for pushing or shunting, so if not secured to the ship at all.

Figure 13.8 Kota Lestari entering the Port of Christchurch, New Zealand.

Berthing the ship

As the master closes the berth, the intent is not to bring the ship directly alongside the berth, but rather parallel to the berth and just short of the berthing position, clear of any forward and aft ships, if present. This is usually performed one ship's breadth between the shoreside and the berth. Once the vessel is stopped off the berth, the master must use the thrusters and tugs to position the vessel into location. Once the ship is alongside, the mooring stations forward and aft throw ashore the spring lines. The spring lines keep the vessel from shifting forwards and aft. Once the spring lines are secured, the headlines and stern lines are thrown ashore and secured. After all the lines are made fast, the winches are set to 40% auto tension and the springs are kept on brake. Where the configuration of the berth is such that long ropes cannot be used, it may be necessary to change the spring lines over to headlines and vice versa. When closing a berth, the master monitors the ship's movement and the distance between the shoreside and the quay, and to other moored ships. To assist the master, ships are fitted with various instruments such as the conning display, voyage management system, and so on, to indicate whether the vessel is moving ahead or astern, the vessel's speed and the amount of set and drift as the ship makes sideways.

Notes

1 Ministry of Transport, China; https://www.cnss.com.cn. Accessed 22 September 2021.
2 This might include crew changes, supplies, and victuals, bunkerage, surveys and inspections, etc.
3 *Bow slamming* refers to the fore of the ship violently striking the water surface due to large waves that can cause structural deformations and failure. *Green water* is the technical term for any water that is present on the upper decks of a vessel due to partial flooding because of the natural motions of the waterbody.
4 Compared to stationary floating structures, such as oil rigs, offshore platforms and FPSO units.
5 An anchor rode encompasses the chain, rope and all the shackles and connectors. The typical rode is measured at approximately eight metres of rode for every one metre of depth. This is referred to as the scope.
6 A short length of anchor cable set between the anchor crown 'D' shackle and the first joining shackle of the cable. The length may consist of a few links which may or may not contain a swivel fitting.
7 The bosun's chair is a seat suspended from the ship and is used to perform maintenance work outside the ship's hull.
8 Oil Companies International Marine Forum (OCIMF).
9 For example, Liquid Natural Gas (LNG) and Liquid Petroleum Gas (LPG).
10 Although on most tugs the towing hook is positioned centrally, when combined with a gob line, the point of attachment can be moved further towards the aft such that the distance is reduced from its original value to 0.45 times the length, waterline (LWL). This helps reduce the amount of manoeuvring the tub needs to perform.
11 Girting is caused when high athwartships towing forces lead a tug to be pulled sideways through the water by the towline. If the tug is unable to manoeuvre out of this position, it is likely to capsize.

Index